# Highrise Office Building Fire
# One Meridian Plaza
# Philadelphia, Pennsylvania

Report by:    J. Gordon Routley
                    Charles Jennings
                    Mark Chubb

This is Report 049 of the Major Fires Investigation Project conducted by TriData Corporation under contract EMW-90-C-3338 to the United States Fire Administration, Federal Emergency Management Agency.

Department of Homeland Security
United States Fire Administration
National Fire Data Center

# U.S. Fire Administration Fire Investigations Program

The U.S. Fire Administration develops reports on selected major fires throughout the country. The fires usually involve multiple deaths or a large loss of property. But the primary criterion for deciding to do a report is whether it will result in significant "lessons learned." In some cases these lessons bring to light new knowledge about fire--the effect of building construction or contents, human behavior in fire, etc. In other cases, the lessons are not new but are serious enough to highlight once again, with yet another fire tragedy report. In some cases, special reports are developed to discuss events, drills, or new technologies which are of interest to the fire service.

The reports are sent to fire magazines and are distributed at National and Regional fire meetings. The International Association of Fire Chiefs assists the USFA in disseminating the findings throughout the fire service. On a continuing basis the reports are available on request from the USFA; announcements of their availability are published widely in fire journals and newsletters.

This body of work provides detailed information on the nature of the fire problem for policymakers who must decide on allocations of resources between fire and other pressing problems, and within the fire service to improve codes and code enforcement, training, public fire education, building technology, and other related areas.

The Fire Administration, which has no regulatory authority, sends an experienced fire investigator into a community after a major incident only after having conferred with the local fire authorities to insure that the assistance and presence of the USFA would be supportive and would in no way interfere with any review of the incident they are themselves conducting. The intent is not to arrive during the event or even immediately after, but rather after the dust settles, so that a complete and objective review of all the important aspects of the incident can be made. Local authorities review the USFA's report while it is in draft. The USFA investigator or team is available to local authorities should they wish to request technical assistance for their own investigation.

This report and its recommendations were developed by USFA staff and by TriData Corporation, Arlington, Virginia, its staff and consultants, who are under contract to assist the USFA in carrying out the Fire Reports Program.

The USFA greatly appreciates the cooperation received from the Philadelphia Fire Department. In particular, the assistance and information provided by Fire Commissioner Roger Ulshafer (ret.), Commissioner Harold Hairston, Deputy Commissioner Christian Scheizer (ret.), Deputy Commissioner Phil McLaughlin, Deputy Commissioner Matthew J. McCrory Jr., Battalion Chief Theodore Bateman, Battalion Chief Richard Bailey, and Lieutenant Matthew Medley were invaluable.

For additional copies of this report write to the U.S. Fire Administration, 16825 South Seton Avenue, Emmitsburg, Maryland 21727. The report is available on the USFA Web site at http://www.usfa.dhs.gov/

# U.S. Fire Administration

## Mission Statement

*As an entity of the Federal Emergency Management Agency (FEMA), the mission of the U.S. Fire Administration (USFA) is to reduce life and economic losses due to fire and related emergencies, through leadership, advocacy, coordination, and support. We serve the Nation independently, in coordination with other Federal agencies, and in partnership with fire protection and emergency service communities. With a commitment to excellence, we provide public education, training, technology, and data initiatives.*

# FOREWORD

This report on the Philadelphia, Pennsylvania, One Meridian Plaza fire documents one of the most significant highrise fires in United States' history. The fire claimed the lives of three Philadelphia firefighters and gutted eight floors of a 38-story fire-resistive building causing an estimated $100 million in direct property loss and an equal or greater loss through business interruption. Litigation resulting from the fire amounts to an estimated $4 billion in civil damage claims. Twenty months after the fire this building, one of Philadelphia's tallest, situated on Penn Square directly across from City Hall, still stood unoccupied and fire-scarred, its structural integrity in question.

This fire is a large scale realization of fire risks that have been identified on many previous occasions. The most significant new information from this fire relates to the vulnerability of the systems that were installed to provide electrical power and to support fire suppression efforts. In this incident there was an early loss of normal electrical power, a failure of the emergency generator and a major problem with the standpipe system, each of which contributed to the final outcome. These experiences should cause responsible individuals and agencies to critically re-examine the adequacy of all emergency systems in major buildings.

When the initial news reports of this fire emerged, attention focused on how a modem, fire-resistive highrise in a major metropolitan city with a well-staffed, well-equipped fire department could be so heavily damaged by fire. The answer is rather simple -- fire departments alone cannot expect or be expected to provide the level of fire protection that modern highrises demand. The protection must be built-in. This fire was finally stopped when it reached a floor where automatic sprinklers had been installed.

This report will demonstrate that the magnitude of this loss is greater than the sum of the individual problems and failures which produced it. Although problems with emergency power systems, standpipe pressure reducing valves, fire alarm systems, exterior fire spread, and building staff response can be identified, the magnitude of this fire was a result of the manner in which these factors interacted with each other. It was the combination of all of these factors that produced the outcome.

At the time of the One Meridian Plaza fire, the three model fire prevention codes had already adopted recommendations or requirements for abating hazards in existing highrise buildings. Each of the model building codes contains explicit requirements for fire protection and means of egress in highrise buildings. Actions were and are underway in many cities and several States to require retrofitting of existing highrise buildings with automatic sprinkler systems, fire detection and alarm systems, and other safety provisions. Since the Meridian Plaza fire, the National Fire Protection Association's Technical Committee on Standpipe Systems has proposed a complete revision of NFPA 14, *Standard for Installation of Standpipe and Hose Systems*. The new version of NFPA 14 was approved by the NFPA membership at the 1992 fall meeting in Dallas, Texas. All of these efforts are necessary and commendable. To prove successful, however, they must take a comprehensive, holistic approach to the problem of highrise fire safety, if we are to keep One Meridian Plaza from being surpassed by yet another devastating fire.

# TABLE OF CONTENTS

*continued on next page*

# Table of Contents (continued)

# Highrise Office Building Fire
## One Meridian Plaza
## Philadelphia, Pennsylvania
## February 23, 1991

Local Contacts:  Commissioner (ret.) Roger Ulshafer
Commissioner Harold Hairston
Deputy Commissioner (ret.) Christian Schweizer
Deputy Commissioner Phil McLaughlin
Deputy Commissioner Matthew J. McCrory, Jr.
Theodore Bateman, Battalion Chief
Richard Bailey, Battalion Chief
Matthew Medley, Lieutenant
City of Philadelphia Fire Department
240 Spring Garden Street
Philadelphia, Pennsylvania 19123-2991
(215) 592-5962

## OVERVIEW

A fire on the 22nd floor of the 38-story Meridian Bank Building, also known as One Meridian Plaza, was reported to the Philadelphia Fire Department on February 23, 1991 at approximately 2040 hours and burned for more than 19 hours. The fire caused three firefighter fatalities and injuries to 24 firefighters. The 12-alarms brought 51 engine companies, 15 ladder companies, 11 specialized units, and over 300 firefighters to the scene. It was the largest highrise office building fire in modern American history -- completely consuming eight floors of the building -- and was controlled only when it reached a floor that was protected by automatic sprinklers. A table summarizing the key aspects of the fire is presented on the following pages.

The fire department arrived to find a well-developed fire on the 22nd floor, with fire dropping down to the 21st floor through a set of convenience stairs. (For an elevation drawing of the building and the 22nd floor plan see Appendix A.) Heavy smoke had already entered the stairways and the floors immediately above the 22nd. Fire attack was hampered by a complete failure of the building's electrical system and by inadequate water pressure, caused in part by improperly set pressure reducing valves on standpipe hose outlets.

# SUMMARY OF KEY ISSUES

| Issues | Comments |
| --- | --- |
| Origin and Cause | The fire started in a vacant 22nd floor office in a pile of linseed oil-soaked rags left by a contractor. |
| Fire Alarm System | The activation of a smoke detector on the 22nd floor was the first notice of a possible fire. Due to incomplete detector coverage, the fire was already well advanced before the detector was activated. |
| Building Staff Response | Building employees did not call the fire department when the alarm was activated. An employee investigating the alarm was trapped when the elevator opened on the fire floor and was rescued when personnel on the ground level activated the manual recall. The fire department was not called until the employee had been rescued. |
| Alarm Monitoring Service | The private service which monitors the fire alarm system did not call the fire department when the alarm was first activated. A call was made to the building to verify that they were aware of the alarm. The building personnel were already checking the alarm at that time. |
| Electrical Systems | Installation of the primary and secondary electrical power risers in a common unprotected enclosure resulted in a complete power failure when the fire damaged conductors shorted to ground. The natural gas powered emergency generator also failed. |
| Fire Barriers | Unprotected penetrations in fire-resistance rated assemblies and the absence of fire dampers in ventilation shafts permitted fire and smoke to spread vertically and horizontally. |
| | Ventilation openings in the stairway enclosures permitted smoke to migrate into the stairways, complicating firefighting. |
| | Unprotected openings in the enclosure walls of 22nd floor electrical closet permitted the fire to impinge on the primary and secondary electrical power risers. |
| Standpipe System and Pressure Reducing Valves (PRVs) | Improperly installed standpipe valves provided inadequate pressure for fire department hose streams using 1-3/4-inch hose and automatic fog nozzles. Pressure reducing valves were installed to limit standpipe outlet discharge pressures to safe levels. The PRVs were set too low to produce effective hose streams; tools and expertise to adjust the valve settings did not become available until too late. |
| Locked Stairway Doors | For security reasons, stairway doors were locked to prevent reentry except on designated floors. (A building code variance had been granted to approve this arrangement.) This compelled firefighters to use forcible entry tactics to gain access from stairways to floor areas. |
| Fire Department Pre-Fire Planning | Only limited pre-fire plan information was available to the Incident Commander. Building owners provided detailed plans as the fire progressed. |
| Firefighter Fatalities | Three firefighters from Engine Company 11 died on the 28th Floor when they became disoriented and ran out of air in their SCBAs. |
| Exterior Fire Spread "Autoexposure" | Exterior vertical fire spread resulted when exterior windows failed. This was a primary means of fire spread. |
| Structural Failures | Fire-resistance rated construction features, particularly floor-ceiling assemblies and shaft enclosures (including stair shafts), failed when exposed to continuous fire of unusual intensity and duration. |
| Interior Fire Suppression Abandoned | After more than 11 hours of uncontrolled fire growth and spread, interior firefighting efforts were abandoned due to the risk of structural collapse. |
| Automatic Sprinklers | The fire was eventually stopped when it reached the fully sprinklered 30th floor. Ten sprinkler heads activated at different points of fire penetration. |

The three firefighters who died were attempting to ventilate the center stair tower. They radioed a request for help stating that they were on the 30th floor. After extensive search and rescue efforts, their bodies were later found on the 28th floor. They had exhausted all of their air supply and could not escape to reach fresh air. At the time of their deaths, the 28th floor was not burning but had an extremely heavy smoke condition.

After the loss of three personnel, hours of unsuccessful attack on the fire, with several floors simultaneously involved in fire, and a risk of structural collapse, the Incident Commander withdrew all personnel from the building due to the uncontrollable risk factors. The fire ultimately spread up to the 30th floor where it was stopped by ten automatic sprinklers.

## THE BUILDING

One Meridian Plaza is a 38-story highrise office building, located at the corner of 15th Street and South Penn Square in the heart of downtown Philadelphia, in an area of highrise and mid-rise structures. On the east side, the building is attached to the 34-story Girard Trust Building and it is surrounded by several other highrise buildings. The front of the building faces City Hall.

One Meridian Plaza has three underground levels, 36 above ground occupiable floors, two mechanical floors (12 and 38), and two rooftop helipads. The building is rectangular in shape, approximately 243 feet in length by 92 feet in width (approximately 22,400 gross square feet), with roughly 17,000 net usable square feet per floor. (See Appendix A for floor plan.) Site work for construction began in 1968, and the building was completed and approved for occupancy in 1973.

Construction was classified by the Philadelphia Department of Licenses and Inspections as equivalent to BOCA Type 1B construction which requires 3-hour fire rated building columns, 2-hour fire rated horizontal beams and floor/ceiling systems, and 1-hour fire rated corridors and tenant separations. Shafts, including stairways, are required to be 2- hour fire rated construction, and roofs must have 1-hour fire rated assemblies.

The building frame is structural steel with concrete floors poured over metal decks. All structural steel and floor assemblies were protected with spray-on fireproofing material. The exterior of the building was covered by granite curtain wall panels with glass windows attached to the perimeter floor girders and spandrels.

The building utilizes a central core design, although one side of the core is adjacent to the south exterior wall. The core area is approximately 38 feet wide by 124 feet long and contains two stairways, four banks of elevators, two HVAC supply duct shafts, bathroom utility chases, and telephone and electrical risers.

### Stairways

The building has three enclosed stairways of concrete masonry construction. Each stairway services all 38 floors. The locations of the two stairways within the building core shift horizontally three or four times between the ground and the 38th floor to accommodate elevator shafts and machine rooms for the four elevator banks. Both of these stairways are equipped with standpipe risers.

Adjacent to the stairway enclosures are separate utility and HVAC shafts. There are pipe and duct penetrations through the shaft and stairway enclosure walls. The penetrations are unprotected around the sleeved pipes and fire dampers are not installed in HVAC ducts penetrating the fire-resistance rated wall assemblies. This effectively creates many openings between the utility shafts, and the individual floors, primarily in the plenum area above the ceilings, as well as between the shafts and the stairway enclosures.

The third enclosed stairway is located at the east end of the building. This stairway attaches the floors of the Meridian Plaza to the corresponding floors of the Girard Trust Building. Adjacent to the east stairway is an additional enclosed utility shaft which also has pipe and duct penetrations through the shaft enclosure walls. There are no fire or smoke barriers around the sleeved pipes and no fire dampers in the HVAC ducts that penetrate the shaft walls.

## Elevators

Elevator service is provided by four zoned elevator banks identified as A through D. Elevator Bank A serves floors 2-11. Elevator Bank B has two shafts which enclose seven elevators: six are passenger elevators that serve floors 12-21, and one is a freight elevator that serves floors 22-38. Elevator Bank C serves floors 21-29, and Elevator Bank D serves floors 29-37. The elevator shafts are constructed of concrete and masonry and extend from the first floor or lower levels to the highest floor served by the individual elevator banks. At the top of each elevator bank is the associated elevator equipment room.

The elevator shafts that serve the upper floors are express rise and do not have openings to the lower floors. Only the Bank C passenger elevators and the freight elevator served the fire floors. The elevator shafts did not appear to play a significant role in the spread of combustion products.

Each elevator lobby is equipped with a smoke detector that, when activated, recalls the elevator cars to the first floor lobby. Firefighter's service (elevator recall) features were added in 1981 under provisions of the State Elevator Code.[1] Occupant use of elevators in emergencies is addressed in the Building Emergency Instructions shown in Appendix B.

## Heating, Ventilation, and Air Conditioning

The heating, ventilation, and air conditioning (HVAC) system is composed of four air handling systems. Two systems are located in the 38th floor mechanical room and service the east and west halves of the upper floors. The other two systems are located in the 12th floor mechanical room and service the east and west halves of the lower floors. Each system supplies air to its respective floors through one or two supply air shafts located within the building core and receives return air from its associated return air shafts. Return air shafts are located at each of the four building corners. Upon examination at selected locations, the HVAC supply and return air shafts did not appear to have fire dampers at the duct penetrations on each floor.

## Plumbing

The bathroom utility piping extends through the 38 floors through pipe chases that are formed by the space between two walls. These pipe chases transfer location as the bathroom locations change floor to floor. Upon a sample examination of the pipe chases, it was found that floor penetrations were not closed or sealed to maintain the integrity of the fire- resistance rated floor/ceiling assemblies.

---

[1] In Pennsylvania, elevators are regulated through the State Department of Labor and Industry.

## Electrical and Communications Risers

The electrical and telephone risers are enclosed in separate rooms on each floor. The rooms are located directly above one another and are intended to function as vertical shafts, with rated separations required at horizontal penetrations from the shafts into floor and ceiling spaces at each level. Within the telephone and electrical rooms, unprotected penetrations of the floor assemblies allow conduits and exposed wires to travel from floor to floor. Several breaches of fire-resistance rated construction were observed in the walls separating the electrical and telephone rooms from the ceiling plenums and occupied spaces on each floor.

## Emergency Power

The building electrical system receives power from two separate electrical substations and is backed-up by an emergency generator. The two sources of power are arranged so that the load would automatically transfer to the second source upon failure of the first. Electrical power for One Meridian Plaza and four adjacent buildings is distributed from the basement of 1414 S. Penn Square.

The electric service enters the building via the basement from the adjoining building and is distributed to the 12th and 38th floor mechanical rooms via the electrical risers in the building core. From the 12th and 38th floor mechanical rooms, electrical power is distributed to the major mechanical systems and to a buss bar riser, which services distribution panels on the individual floors.

Emergency power was provided by a 340 kw natural gas-fired generator located in the 12th floor mechanical room. The generator was sized to supply power for emergency lighting and the fire alarm system, the fire pump located on the 12th floor and one car in each bank of elevators. The generator's fuel was supplied by the building's natural gas service. This generator was not required by the building code, since the building's electrical power was supplied by two separate substations.

The generator was reported to have been tested weekly. The last recorded test date was January 30, almost four weeks before the fire, and the maintenance records indicate that problems were encountered during engine start-up under load conditions at that time. During a detailed inspection following that test, a damaged part was discovered and replaced. After the repair, the generator was started without a load and appeared to work properly, but no subsequent tests were performed to determine if the problems persisted under load conditions.

Records of earlier maintenance and test activity suggest that load tests were performed only occasionally. Test and maintenance records indicate a long history of maintenance problems with the emergency generator system. Many of these problems became manifest during or immediately after conducting tests under load.

## FIRE PROTECTION SYSTEMS

At the time of construction, the Philadelphia Building Code required only a local fire alarm system with manual stations at each exit and smoke detectors in the supply and return air shafts. Hose stations supplied from the domestic water service and portable fire extinguishers were required for occupant use. Dry standpipes were installed for fire department use. Below ground levels were required to be provided with automatic sprinklers.

As a result of local code changes, several improvements to the fire protection systems were made in the years following the building's construction.

In 1981, the Philadelphia Department of Licenses and Inspections implemented amendments to the fire code which were intended to address the life safety of highrise building occupants. These requirements included installation of stair identification signs, provisions to permit stairway re-entry, and installation of smoke detection in common areas in the path of access to exits. The "common areas" provision of the code was intended to address corridors and exit passageways in multi-tenant floors. The smoke detector requirements were interpreted in such a way that single tenant "open plan" floors were only required to have detectors installed at the exits; the entire floor, although open, was not considered a "common area.". Smoke detectors were also installed in the return air plenum adjacent to the return air shaft intakes in each corner of the building. These provisions required that building owners file permits for this work within one year of the code change. City records do not indicate when this work was performed in this particular building or if it was inspected and approved.

## Fire Detection and Alarm Systems

At the time of construction, One Meridian Plaza was equipped with a coded manual fire alarm system with pull stations installed adjacent to each of the three exit stairwells on each floor. Smoke detection was provided in the major supply and return air ducts at the mechanical floor levels.

After the 1981 fire code amendments were enacted, the hardware on stairway doors was required to allow access from stairs back to floor areas or to be unlocked automatically in the event that the fire alarm was activated. One Meridian Plaza was granted a variance from this provision and generally had unlocked doors every three floors.

Approximately one and a half years before the fire, a public address system was installed throughout the building. This system was operable from the lobby desk and had the capability of addressing floors, stairways, elevator machine rooms, and elevators. Two-way communication was possible with elevators and elevator machine rooms.

As additional devices and systems were installed, they were connected to the fire alarm system to sound through the single-stroke bells originally installed with the manual fire alarm system. Smoke detector and water flow signals were assigned their own codes to allow annunciation not only at the lobby but throughout the building for those members of the building staff who knew the codes.

## Standpipes

The occupant use standpipe system, which was connected to the domestic water supply, provided two outlets per floor with 100 feet of 1-1/2-inch hose and a nozzle. The hose cabinets were located in corridors on each floor.

A dry standpipe system was originally installed with 6 inch risers in the west and center stair towers and outlets for 2-1/2 fire department hoselines at each floor level. This system was converted to a wet riser system in 1988, to supply automatic sprinklers on some of the upper floors. An 8 inch water supply was provided to deliver water to two 750 gpm electric fire pumps, one in the basement and one on the 12th floor.

The basement pump supplied the lower standpipe zone (floors B-12) while the 12th floor pump served the upper zone (floors 13-38).

There was no standpipe in the east stair tower.

A November 1988 Board of Building Standards decision permitted both zones to be served by a common fire department connection, as part of a plan that would provide for the installation of automatic sprinklers on all floors by November 1993.[2]

Due to the height of the zones and the installation of fire pumps, pressures exceeded the 100 psi limit permitted by NFPA 14, *Installation of Standpipe and Hose Systems* at the standpipe hose outlets on several lower floors in each zone. Pressure restricting devices, which limit the discharge through standpipe outlets by restricting the orifice, were installed on the mezzanine and second floor levels and on floors 26 through 30. Pressure reducing valves, which regulate both static pressure and discharge pressure under variable flow conditions, were installed on floors 13 through 25.

Both types of devices prevent dangerous discharge pressures from hose outlets at the lower floors of each standpipe zone. The Philadelphia Fire Department investigators report that the plans submitted at the time the standpipes were converted did not indicate that PRVs were to be installed.

## Automatic Sprinklers

Only the service floors located below grade were protected by automatic sprinklers at the time of construction. Conversion of the dry standpipe to a wet system with fire pumps facilitated the installation of automatic sprinklers throughout the building. At the request of selected tenants, sprinklers were installed on several floors during renovations, including all of the 30th, 31st, 34th, and 35th floors, and parts of floors 11 and 15. Limited service sprinklers, connected to the domestic water supply system, were installed in part of the 37th floor. The building owners had plans to install sprinklers on additional floors as they were renovated.

# THE FIRE

## Delayed Report

At approximately 2023 hours on February 23, 1991, a smoke detector was activated on the 22nd floor of the One Meridian Plaza building. The activated detector is believed to have been located at the entrance to the return air shaft in the northeast corner of the building. At that time there were three people in the building, an engineer and two security guards.[3] The alarm sounded throughout the building and elevator cars automatically returned to the lobby. The building engineer investigated the alarm using an elevator on manual control to go to the 22nd floor. The central station monitoring company that served the building reportedly called the guard desk in the lobby to report the alarm. The call came in before the engineer reached the fire floor, and the alarm company was told that the source of the alarm was being investigated. The alarm company did not notify the fire department at that time.

When the elevator doors opened at the 22nd floor, the engineer encountered heavy smoke and heat. Unable to reach the buttons or to leave the elevator car to seek an exit, the building engineer became trapped. He was able to use his portable radio to call the security guard at the lobby desk requesting

---

[2] Philadelphia Fire Department, "Investigative Report," Section M, p. 2.

[3] The building staff regulated the after-hours population of the building through a lighting request system where tenants lights would be turned on for the duration of their work. In addition, there was a security system in the building that recorded any passage through stairwell doors.

assistance. Following the trapped engineer's instructions, the security guard in the lobby recalled the elevator to the ground floor using the Phase II firefighter's safety feature.

The second security guard monitored the radio transmissions while taking a break on the 30th floor. This guard initially mistook the fire alarm for a security alarm believing that he had activated a tenant's security system while making his rounds. He evacuated the building via the stairs when he heard the building engineer confirm there was a fire on the 22nd floor.

The roving guard reported that as he descended from the 30th floor the stairway was filling with smoke. He reached the ground level and met the engineer and the other security guard on the street in front of the building.

The Philadelphia Fire Department report on the incident states that the lobby guard called the alarm monitoring service to confirm that there was an actual fire in the building when the engineer radioed to her from the 22nd floor. After meeting outside and accounting for each other's whereabouts the three building personnel realized that they had not yet called the fire department.

The first call received by the Philadelphia Fire Department came from a passerby who used a pay telephone near the building to call 9-1-1. The caller reported smoke coming from a large building but was unable to provide the exact address. While this call was still in progress, at approximately 2027 hours, a call was received from the alarm monitoring service reporting a fire alarm at One Meridian Plaza.

## Initial Response

The Philadelphia Fire Department dispatched the first alarm at 2027 hours consisting of four engine and two ladder companies with two battalion chiefs. The first arriving unit, Engine 43, reported heavy smoke with fire showing from one window at approximately the mid-section of the building at 2031 hours. A security guard told the first arriving battalion chief that the fire was on the 22nd floor. Battalion Chief 5 ordered a second alarm at 2033 hours.

While one battalion chief assumed command of the incident at the lobby level, the other battalion chief organized an attack team to go up to the fire floor. (The Philadelphia Fire Department's "Highrise Emergency Procedures" Operation Procedure 33 is presented in Appendix C.) The battalion chief directed the attack team to take the low-rise elevators up the 11th floor and walk up from there.

## Electrical Power Failure

Shortly after the battalion chief and the attack team reached the 11th floor there was a total loss of electrical power in the building. This resulted when intense heat from the fire floor penetrated the electrical room enclosure. The heat caused the cable insulation to melt resulting in a dead short between the conductor and the conduit in both the primary and secondary power feeds, and the loss of both commercial power sources. The emergency generator should have activated automatically, but it failed to produce electric power. These events left the entire building without electricity for the duration of the incident in spite of several efforts to restore commercial power and to obtain power from the generator.

This total power failure had a major impact on the firefighting operations. The lack of lighting made it necessary for firefighters to carry out suppression operations in complete darkness using only battery powered lights. Since there was no power to operate elevators, firefighters were forced to hand carry all suppression equipment including SCBA replacement cylinders up the stairs to the staging area that was established on the 20th floor. In addition, personnel had to climb at least 20 floors to relieve fellow firefighters and attack crews increasing the time required for relief forces to arrive. This was a problem for the duration of the incident as each relief crew was already tired from the long climb before they could take over suppression duties from the crews that were previously committed.

## Initial Attack

As the initial attack crews made their way toward the 22nd floor they began to encounter smoke in the stairway. At the 22nd floor they found the west stair tower door locked. The door was already warped and blistering from the heat, and heavy fire could be seen through the door's wire glass window. A 1-3/4-inch handline was stretched up the stairway from a standpipe connection on the floor below and operated through the window while a ladder company worked on forcing open the door.

It took several minutes before the door could be forced open and an attempt could be made to advance onto the fire floor with the 1-3/4-inch attack line. The crews were not able to penetrate onto the 22nd floor due to the intense heat and low water pressure they were able to obtain from their hoseline.

An entry was also made on the 21st floor where the firefighters were able to see fire on the floor above through the open convenience stair. They attempted to use an occupant hoseline to attack the fire but could not obtain water from that outlet. They then connected a 1-3/4 inch attack line to the standpipe outlet in the stairway, but they could not obtain sufficient pressure to attack the flames. A Tactical Command Post was established on the 21st floor and a staging area was set up on floor 20.

## Fire Development

By this time fire was visible from several windows on the 22nd floor and crews outside were evacuating the area around the building and hooking up supply lines to the building's standpipe connections. As flames broke through several more windows around a major portion of the fire floor, the floor above was subject to autoexposure from flames lapping up the side of the building. Additional alarms were called to bring personnel and equipment to the scene for a large scale fire suppression operation.

As the fire developed on the 22nd floor, smoke, heat, and toxic gases began moving through the building. Vertical fire extension resulted from unprotected openings in floor and shaft assemblies, failure of fire-resistance rated floor assemblies, and the lapping of flames through windows on the outside of the building.

## Water Supply Problems

The normal attack hoselines used by the Philadelphia Fire Department incorporate 1-3/4-inch hoselines with automatic fog nozzles designed to provide variable gallonage at 100 psi nozzle pressure. The pressure reducing valves in the standpipe outlets provided less than 60 psi discharge pressure, which was insufficient to develop effective fire streams. The pressure reducing values (PRVs) were field adjustable using a special tool. However, not until several hours into the fire did a technician knowledgeable in the adjustment technique arrive at the fire scene and adjust the pressure on several of the PRVs in the stairways.

When the PRVs were originally installed, the pressure settings were improperly adjusted. Index values marked on the valves did not correspond directly to discharge pressures. To perform adjustments the factory and field personnel had to refer to tables in printed installation instructions to determine the proper setting for each floor level.[4] For more detailed information about PRVs see Appendices D and E.

Several fire department pumpers were connected to the fire department connections to the standpipe system in an attempt to increase the water pressure. The improperly set PRVs effectively prevented the increased pressure in the standpipes from being discharged through the valves. The limited water supply prevented significant progress in fighting the fire and limited interior forces to operating from defensive positions in the stairwells. During the next hour the fire spread to the 23rd and 24th floors primarily through autoexposure, while firefighters were unable to make entry onto these floors due to deteriorating heat and smoke conditions and the lack of water pressure in their hoselines. Windows on the 22nd floor broke out and the 23rd and 24th floor windows were subject to autoexposure from flames lapping up the sides of the building.

On the street below pedestrians were cleared from the area because of falling glass and debris as more and more windows were broken out by the fire. Additional hoselines were connected to the standpipe connections, attempting to boost the water pressure in the system. However, the design of the PRVs did not allow the higher pressures to reach the interior hose streams. Additional alarms were requested to bring a five-alarm assignment to the scene.

## Three Firefighters Lost

As firefighters attempted to make entry to the burning floors from the stairways, heavy smoke continued to build up within the stair shafts and banked down from the upper floors. An engine company was assigned to attempt to open a door or hatch to ventilate the stairways at the roof level to allow the smoke and heat to escape. A captain and two firefighters from Engine 11 started up the center stair from the 22nd floor with this assignment. Engine 11 subsequently radioed that they had left the stairway and were disoriented in heavy smoke on the 30th floor. Attempts were made to direct the crew by radio to one of the other stairways.

Shortly thereafter a radio message was received at the Command Post from Engine 11's Captain requesting permission to break a window for ventilation. This was followed moments later by a message from a crew member of Engine 11 reporting that "the captain is down." Approval was given to break the window and rescue efforts were initiated to search for the crew. Search teams were sent from below and a helicopter was requested to land a team on the roof. The search teams were able to reach the 30th floor, which was enveloped in heavy smoke, but were unable to find the missing firefighters. They then searched the floors above without success. An eight member search team became disoriented and ran out of air in the mechanical area on the 38th floor, while trying to find an exit to the roof. They were rescued by the team that had landed on the roof and were transported back to ground level by the helicopter.

---

[4] The pressure reducing valves in the vicinity of the fire floor (floors 18 through 20) were set at "80" on the valve index which corresponded to a discharge pressure between 55 and 57 psi, depending on the elevation. This would provide a nozzle pressure of 40 to 45 psi at the end of a 150- to 200-foot hoseline.

Several attempts were made to continue the search, until helicopter operations on the rooftop had to be suspended due to the poor visibility and the thermal drafts caused by the heat of the fire. The helicopter crew then attempted an exterior search, using the helicopter's searchlight, and at 0117 located a broken window on the southeast corner of the 28th floor, in an area that could not be seen from any of the surrounding streets. Another rescue team was assembled and finally located the three missing members just inside the broken window on the 28th floor at approximately 0215. At that time the fire was burning on the 24th and 25th floors and extending to the 26th.

The victims were removed to the Medical Triage Area on the 20th floor, but resuscitation efforts were unsuccessful and they were pronounced dead at the scene. An estimated three to four hours had elapsed since they had reported that they were in trouble and all had succumbed to smoke inhalation.[5]

The three deceased members of Engine Company 11 were Captain David P. Holcombe (age 52), Firefighter Phyllis McAllister (43), and Firefighter James A. Chappell (29).

Prior to being assigned to this task, the crew had walked up to the fire area wearing their full protective clothing and SCBA and carrying extra equipment. It is believed that they started out with full SCBA cylinders, but it is not known if they became disoriented from the heavy smoke in the stairway, encountered trouble with heat build-up, or were exhausted by the effort of climbing 28 floors. Some combination of these factors could have caused their predicament. Unfortunately, even after breaking the window they did not find relief from the smoke conditions which were extremely heavy in that part of the building.

## Continuing Efforts to Improve Water Supply

Because of the difficulty in obtaining an adequate water supply, a decision was made to stretch 5-inch lines up the stairs to supply interior attack lines. The first line was stretched up the west (#1) stairwell to the 24th floor level and was operational by 0215, approximately six hours into the fire. At 0221, a 12th alarm was sounded to stretch a second line, in the center (#2) stair. At 0455, a third 5-inch line was ordered stretched, in the east (#3) stair. The operation in the east stair was discontinued at the 17th floor level at 0600. While the 5-inch lines were being stretched, a sprinkler contractor arrived at the scene and began manually adjusting the pressure reducing valves on the standpipe connections. This improved the discharge pressure in the hoses supplied by the standpipe system, finally providing normal handline streams for the interior fire suppression crews. At this point, however, the fire involved several floors and could not be contained with manual hose streams.

## Firefighting Operations Suspended

All interior firefighting efforts were halted after almost 11 hours of uninterrupted fire in the building. Consultation with a structural engineer and structural damage observed by units operating in the building led to the belief that there was a possibility of a pancake structural collapse of the fire damaged floors. Bearing this risk in mind along with the loss of three personnel and the lack of

---

[5]The exact time that Engine 11 was assigned to attempt ventilation and the time the crew reported they were in trouble are not known, since the tactical radio channel they were using is not recorded and detailed time records of this event were not maintained during the incident. Estimates from individuals who were involved suggest that the assignment was made between 2130 and 2200 hours and search efforts were initiated between 2200 and 2230 hours. The bodies were located at approximately 0215 hours.

progress against the fire despite having secured adequate water pressure and flow for interior fire streams, an order was given to evacuate the building at 0700 on February 24. At the time of the evacuation, the fire appeared to be under control on the 22nd though 24th floors. It continued to burn on floors 25 and 26 and was spreading upward. There was a heavy smoke condition throughout most of the upper floors. The evacuation was completed by 0730.

After evacuating the building, portable master streams directed at the fire building from several exposures, including the Girard Building #1 and One Centre Plaza, across the street to the west were the only firefighting efforts left in place.

## Fire Stopped

The fire was stopped when it reached the 30th floor, which was protected by automatic sprinklers. As the fire ignited in different points this floor level through the floor assembly and by autoexposure through the windows, 10 sprinkler heads activated and the fires were extinguished at each point of penetration. The vertical spread of the fire was stopped solely by the action of the automatic sprinkler system, which was being supplied by fire department pumpers. The 30th floor was not heavily damaged by fire, and most contents were salvageable. The fire was declared under control at 3:01 p.m., February 24, 1991.

## ANALYSIS

### Smoke Movement

The heated products of combustion from a fire have a natural buoyancy, which causes them to accumulate in the upper levels of a structure. In a highrise building the stairways, elevator shafts, and utility shafts are natural paths for the upward migration of heated products of combustion.

Stack effect is a natural phenomenon affecting air movement in tall buildings. It is characterized by a draft from the lower levels to the upper levels, with the magnitude of the draft influenced by the height of the building, the degree of air-tightness of exterior walls of the building, and temperature differential between inside and outside air.[6] This effect was particularly strong on the night of the fire due to the cold outside temperature. Interior air leakage rates, through shaft walls and openings, also modulate the rate of air flow due to stack effect. Smoke and toxic gases become entrained in this normal air movement during a fire and are carried upward, entering shafts through loose building construction or pipe and duct penetrations. The air flow carries smoke and gases to the upper portions of the structure where the leakage is outward.

At the upper portions of the structure, smoke and toxic gases fill the floors from the top floor down toward the fire, creating a dangerous environment for building occupants and firefighters. During the investigation of this fire, this upward flow was evidenced by the presence of heavy soot in the 38th floor mechanical room and all the upper floors of the building. The path of smoke travel to the upper floors was vividly evidenced by the soot remnants in HVAC shafts, utility chases, return air shafts, and exhaust ducts.

---

[6] Fitzgerald, R. (1981), "Smoke Movement in Buildings," in Fire Protection Handbook, 15th ed., McKinnon, G. P., ed., Quincy, MA: National Fire Protection Association, p. 3-32.

## Fuel Loading

Fuel loading on the fire floors consisted mainly of files and papers associated with securities trading and management consulting. At least one floor had a significant load of computer and electronic equipment. In some cases, correlation could be found between heavy fuel load and damage to structural members in the affected area. From the 22nd floor to the 29th floor, the fire consumed all available combustible materials with the exception of a small area at the east end of the 24th floor.

## Structural Conditions Observed

Prior to deciding to evacuate the building, firefighters noticed significant structural displacement occurring in the stair enclosures. A command officer indicated that cracks large enough to place a man's fist through developed at one point. One of the granite exterior wall panels on the east stair enclosure was dislodged by the thermal expansion of the steel framing behind it. After the fire, there was evident significant structural damage to horizontal steel members and floor sections on most of the fire damaged floors. Beams and girders sagged and twisted--some as much as three feet--under severe fire exposures, and fissures developed in the reinforced concrete floor assemblies in many places. Despite this extraordinary exposure, the columns continued to support their loads without obvious damage.

## INCIDENT COMMAND

During nearly 19 hours of firefighting, the Philadelphia Fire Department committed approximately 316 personnel operating 51 engine companies, 15 ladder companies, and 11 specialized units, including EMS units, to the 12-alarm incident. The incident was managed by 11 battalion chiefs and 15 additional chief officers under the overall command of the Fire Commissioner. All apparatus and personnel were supplied without requesting mutual aid. Off-duty personnel were recalled to staff reserve companies to maintain protection for all areas of the city. Philadelphia uses an incident management system known within the department as Philadelphia Incident Command (PlC). It is based on the ICS system taught at the National Fire Academy.

## Operations

The Department's standard operating procedures for a highrise incident were implemented at the time of arrival. Incident commanders were confronted with multiple simultaneous systems failures. As a result, command and control decisions were based on the need to innovate and to find alternate approaches to compensate for the normal systems that firefighters would have relied on to bring this incident to a more successful conclusion.

Philadelphia Fire Department tactical priorities in a highrise fire focus on locating and evacuating exposed occupants and making an aggressive interior attack to confine the fire to the area or at least the floor of origin. Confronted with total darkness, impaired vertical mobility because the elevators were inoperable, water supply deficiencies which made initial attack efforts ineffective, vertical fire spread via unprotected interior openings and external auto-exposure, and worsening heat and smoke conditions in the stairways, the tactical focus shifted to finding something (perhaps anything) which could be accomplished safely and effectively.

When Engine 11's crew reported their predicament, the priority changed to attempting to locate and rescue the trapped firefighters. Unfortunately, these efforts were in vain and nearly proved tragic when the eight firefighters conducting search and rescue operations became disoriented and ran out of air in the 38th floor mechanical room and nearly perished while trying to locate a roof exit. The rescue of these members was extremely fortunate in a situation that could have resulted in an even greater tragedy.

## Communications

As is often the case, communication at such a large incident presented a serious challenge to maintaining effective command and control. The loss of electrical power plunged the entire building into total darkness, forcing firefighters to rely on portable lights. This impacted even face-to-face communications by making it difficult for people to identify with whom they were talking.

Radio communication was also affected by the significant duration of the incident. A field communications van was brought to the scene early in the incident with a supply of spare radios and batteries, but it was a major challenge to provide charged batteries for all of the radios that were in use in the incident.

To ease congestion on fireground radio channels, cellular telephones were used to communicate between the Command Post in the lobby and the staging area on the 20th floor. Several other communications functions took advantage of the cellular telephone capability.

## Logistics

The Logistics Section was responsible for several functions including refilling SCBA cylinders, supplying charged radio batteries, and stretching the 5-inch supply line up the stairways. These were monumental endeavors which required the labor of approximately 100 firefighters. Equipment and supplies were in constant demand including handlights and portable lighting, deluge sets, hose, nozzles and other equipment. The Staging Area on the 20th floor included the Medical and Rehabilitation sectors.

The Philadelphia Fire Department used its highrise air supply system to refill air cylinders on the 20th floor. Falling glass and debris severed the airline, which is extended from the air compressor vehicle outside the building to the staging area, and the system had to be repaired and reconnected at the scene.

## Safety

When things go wrong on a scale as large as One Meridian Plaza, safety becomes an overriding concern. Firefighters were continually confronted with unusual danger caused by multiple system failures during this incident. The deaths of the three firefighters and the critical situation faced by the rescue team that was searching for them are clear evidence of the danger level and the difficulties of managing operations in a dark, smoke-filled highrise building.

A perimeter was set up around the building to prevent injuries from falling glass and stone panels, but it was necessary for personnel to cross this zone to enter and exit the building and to maintain the hoselines connected to the standpipe system. One firefighter was seriously injured when struck by falling debris while tending the hoselines. In addition, all supplies and equipment needed inside the building had to cross the safety perimeter at some point.

Many firefighters working inside the building were treated for minor injuries and fatigue during the fire. Rest and rehabilitation sectors contributed to firefighter safety by improving mental and physical stamina, and a medical triage treatment area was established on the 20th floor.

The physical and mental demands on personnel were extraordinary. In addition to managing the physical safety of personnel, their emotional and psychological well-being were considered. The department activated its critical incident stress debriefing program and relieved first and second alarm companies soon after discovering that the crew of Engine 11 had died on the 28th floor. More than 90 firefighters were debriefed on site after the dead firefighters were evacuated. The CISD involvement continued after the fire, due to the tremendous impact of the loss and the risk to the hundreds of firefighters who were involved in the incident.

The most courageous safety decision occurred when Fire Commissioner Roger Ulshafer ordered the cessation of interior firefighting efforts and evacuated the building due to the danger of structural collapse. Firefighters did not reenter the structure until the fire had been controlled by the automatic sprinklers on the 30th floor and burned out all of the available fuels on the fire-involved lower levels.

## BUILDING AND FIRE REGULATIONS

When One Meridian Plaza construction began in 1968, the City of Philadelphia was enforcing the 1949 edition of the Philadelphia Building Code. This code was of local origin and contained minor amendments that had been incorporated since its enactment. This building code made no distinction between highrise and other buildings; and therefore, no special highrise construction features were required. The general focus of the code was to provide a high degree of fire-resistive construction rather than relying on automatic sprinkler protection or compartmentation.

In 1984, Philadelphia switched from a locally developed building code to one based on the BOCA Basic Building Code/1981. That code has since been updated, and the BOCA National Building Code/1990 is currently in force in Philadelphia. Both of these codes contain provisions expressly addressing highrise building fire protection, including a requirement for automatic sprinkler systems in all new highrise buildings.

As a result of this fire, the City of Philadelphia has adopted an ordinance requiring all existing highrise buildings to be protected by automatic sprinklers by 1997. Also, officials of the Philadelphia Fire Department have discussed proposing adoption of the BOCA National Fire Prevention Code with local amendments, as opposed to continuing to develop their Philadelphia Fire Prevention Code locally.

In 1981, the City enacted amendments to its Fire Code requiring the installation of special fire protection features in existing highrise buildings. These modifications included:

- Fire alarm systems with smoke detection in elevator lobbies, entrances to exit stairways, return air plenums, corridors, and other common or public areas.

- Stairway identification signs, (i.e., identification of the stairway, floor level, and the top and bottom levels of the stairway).

- Stairway re-entry to permit occupants to retreat from stairways compromised by smoke or fire and return to tenant spaces. (In the event doors were locked from the stairway side for security reasons, provisions had to be made to unlock doors automatically upon activation of the fire alarm system.)

In November 1984, the Philadelphia Department of Licenses and Inspections issued a notice of violation to the owners of One Meridian Plaza requiring compliance with these amendments. The Board of Safety and Fire Prevention later granted the owners a variance to permit stairway doors to be locked, provided that doors were unlocked to permit re-entry every third floor.[7]

## FIRE CODE ENFORCEMENT

The preparation and adoption of fire safety regulations is managed by the Philadelphia Fire Department under the direction of the fire marshal. However, the department does not perform or direct compliance inspections of individual properties. Fire code enforcement is delegated to the Department of Licenses and Inspections (L&I) by city charter. This department performs the functions of the building official in Philadelphia.

The fire department conducts inspections of properties applying for variances, follows-up citizen complaints, and makes referrals to L&I. The block inspection program detailed in Fire Department Operational Procedure 4 (see Appendix F) provides for the annual inspection of all buildings except one and two family dwellings. However, fire department activities to detect and abate hazards are primarily of an educational nature. Guidelines for referring serious or continuing hazards to L&I are not detailed in the Block Inspection procedure; however, information regarding the maintenance of referral and appeal records for individual properties is detailed.

It has been questioned whether the working relationship between line company personnel, the fire marshal's office, and the Department of Licenses and Inspections produces effective fire code compliance. Senior fire department officials have expressed considerable dissatisfaction with the relationship between the fire department and L&I, and continue to advocate a more active role for the fire department in code enforcement matters.

Fire inspection records for One Meridian Plaza were examined after the fire to document code enforcement actions requiring the installation or upgrade of fire protection features required by the 1981 fire code amendments. An August 17, 1990, L&I violation notice cited the owner for failing to pay a non-residential inspection fee and noted that a reinspection would be conducted within 30 days. However, no record of a subsequent inspection was produced.

## LESSONS LEARNED

Perhaps the most striking lesson to be learned from the One Meridian Plaza highrise fire is what can happen when everything goes wrong. Major failures occurred in nearly all fire protection systems. Each of these failures helped produce a disaster. The responsibility for allowing these circumstances to transpire can be widely shared, even by those not directly associated with the events on and before February 23, 1991.

To prevent another disaster like One Meridian Plaza requires learning the lessons it can provide. The consequences of this incident are already being felt throughout the fire protection community. Major code changes have already been enacted in Philadelphia (see Appendix G) and new proposals

---

[7] Records and reports provided by the Philadelphia Fire Department during this investigation do not indicate the dates of either the appeal or this variance. Firefighters did report having to force entry on several floors during firefighting because some stairway doors were locked.

are under consideration by the model code organizations. These changes may eventually reduce the likelihood of such a disaster in many communities.

### 1. Automatic sprinklers should be the standard level of protection in highrise buildings.

The property conservation and life safety record of sprinklers is exemplary, particularly in high-rise buildings. While other fire protection features have demonstrated their effectiveness over time in limiting losses to life and property, automatic sprinklers have proven to provide superior protection and the highest reliability. Buildings in some of the nation's largest cities, designed and built around effective compartmentation, have demonstrated varying success at containing fires, but their effectiveness is often comprised by inadequate design or installation and may not be effectively maintained for the life of the building. Even with effective compartmentation, a significant fire may endanger occupants and require a major commitment of fire suppression personnel and equipment. Retrofitting of automatic sprinklers in existing buildings has proven effective in taking the place of other systems that have been found to be inadequate.

### 2. Requirements for the installation of automatic sprinklers are justified by concerns about firefighter safety and public protection effectiveness, as well as traditional measures such as life safety and property conservation.

The property protection value of sprinklers was recognized long before life safety became a popular justification for installing fire protection. Life safety has become the primary concern in recent times, justifying the installation of automatic sprinklers in highrise buildings. The value of sprinklers as a means of protecting firefighters has rarely been discussed.

Members of the fire service should promote automatic fire sprinklers if for no other reason than to protect themselves. Requiring the installation and maintenance of built-in fire protection should become a life safety issue for firefighters.[8] The opposition to retrofit protection has consistently cited cost concerns. Communities need to be made aware that the costs they defer may be paid by firefighters in terms of their safety. This is above and beyond the potential loss to citizens and businesses that is usually considered.

### 3. Code assumptions about fire department standpipe tactics proved invalid.

One of the principal code assumptions affecting fire department operations at One Meridian Plaza concerned the installation of standpipe pressure reducing valves. The rationale for PRVs is the concern that firefighters would be exposed to dangerous operating pressures and forces if they connected hoselines to outlets near the base of standpipe risers of substantial height, particularly those supplied by stationary fire pumps. *For example, in a 275 foot high standpipe zone (the highest permitted using standard pipe and fittings), a pressure of 184 psi is required at the base of the riser to overcome elevation and produce the minimum required outlet pressure of 65 psi at the top of the riser. At this pressure, a standard 2-1/2-inch fire hose fitted with a 1-1/8-inch straight bore nozzle would produce a back pressure (reaction force) in excess of 500 pounds.* This is a well-founded concern; however, it is built upon the assumption that fire departments use 2-1/2-inch attack lines and straight bore nozzles to attack fires from standpipes. Most fire departments today use 1-3/4-inch and 2-inch hose with fog nozzles

---

[8] A study by Charles Jennings reports that the firefighter injury rate in non-sprinklered highrise buildings is seven times higher than in comparable buildings equipped with automatic sprinklers. "In Highrise Fires, Sprinklers Beat Compartmentation —Hands Down." *U.S. Fire Sprinkler Reporter*, April 1992, pp. 1,5-7.

for interior attack. These appliances require substantially greater working pressures to achieve effective hose streams.

In the aftermath of this incident, the NFPA Technical Committee on Standpipes has proposed a complete revision of NFPA 14[9] to more closely reflect current fire department operating practices. Section 5-7 of the proposed standard requires a minimum residual pressure of 150 psi at the required flow rate from the topmost 2-1/2-inch hose outlet and 65 psi at the topmost 1-1/2-inch outlet (presumably for occupant use). Minimum flow rates of 500 gpm for the first standpipe and 250 gpm for each additional standpipe remain consistent with past editions of the standard. The proposed new requirements limit the installation of pressure regulating devices to situations where static pressures at hose outlets exceed 100 psi for occupant use hose or 175 psi for fire department use hose. This will provide substantially greater flow and pressure margins for fire department operations. These requirements are intended to apply to new installations and are not retroactive.

---

Firefighters at One Meridian Plaza had great difficulty determining how to improve flow and pressure from hose outlets during the fire. Even if firefighters could have closely examined the valves, with good light and under less stressful conditions, it is unlikely that they would have been able to readjust the valves. Numerical indicators on the valve stems represented an index for adjustment not the actual discharge pressure. (This may have confused the technicians responsible for installing and maintaining the valves. Investigators found valves set at "20" and "80" on the index markings. To achieve 65 psi would have required a setting from 88 to 91 on the index. A setting of 150 to 158 was necessary to produce the maximum allowable 100 psi.)

Pressure regulating devices come in three different types:

- **Pressure restricting devices** which reduce pressure under flowing conditions by reducing the cross- sectional area of the hose outlet.

- **Pressure control valves** are pilot-operated devices which use water pressure within the system to modulate the position of a spring-loaded diaphragm within the valve to reduce downstream pressure under flowing and non-flowing conditions.

- **Pressure reducing valves** use a spring-loaded valve assembly to modulate the position of the valve disc in the waterway to reduce the downstream pressure under flowing and non-flowing conditions.

Further differentiation within each of these types results from differences in manufacturer specifications. (Details are provided in the Philadelphia Fire Department fact sheet on pressure regulating devices in Appendix G.) Some devices are field adjustable, some are not. Some can be removed to permit full, unrestricted flow, others cannot.

---

[9]The report of the Technical Committee on Standpipes appears in the NFPA 1992 *Fall Meeting Technical Committee Reports,* pp 331-367.

4. **The requirements and procedures for design, installation, inspection, testing, and maintenance of standpipes and pressure reducing valves must be examined carefully.**

The proposed revision of NFPA 14 (1993) and a new NFPA document, NFPA 25, *Standard for the Installation, Testing, and Maintenance of Water-Based Fire Protection Systems* (1992), address many of the concerns arising from this fire regarding installation and adjustment of pressure reducing valves. NFPA 14 requires acceptance tests to verify proper installation and adjustment of these devices. NFPA 25 requires flow tests at five year intervals to verify proper installation and adjustment.

Neither of these standards proposes changes in the performance standards for the design of pressure reducing valves.

Standard performance criteria for the design and operation of each type of valve should be adopted to encourage user-friendly designs that will permit firefighters to achieve higher pressure and flow rates without interrupting firefighting operations. The operation and adjustment of valves should be easy to identify and clearly understandable by inspection and maintenance personnel without reliance on detailed operating or maintenance instructions.

It is extremely important to have all systems and devices thoroughly inspected and tested at the time of installation and retested on a regular basis. Fire suppression companies that respond to a building should be familiar with equipment that is installed in its fire protection systems and confident that it will perform properly when needed.

5. **Inconsistencies between code assumptions and firefighting tactics must be addressed.**

The inconsistency between fire department tactics and design criteria for standpipe hose outlet pressures was widely recognized before this fire. However, little was done to change fire department tactics or to amend the code requirements for standpipe installations.

Fire departments utilize lightweight hose and automatic nozzles for the same reasons the code requires pressure reducing valves: firefighter safety. The inconsistency between these approaches can cause serious problems. Where pressure reducing valves are not installed, fire departments can usually augment water supplies by connecting to the fire department connections. However, when contemporary firefighting tactics are employed and improperly adjusted PRVs are installed, the combination is likely to produce hose streams with little reach or effectiveness.

The PRV equipped hose outlets on the 22nd floor of One Meridian Plaza, adjusted as reported at the time of the fire, would have produced nozzle pressures of approximately 40 psi. This is insufficient for a straight stream device and dangerously inadequate for a fog nozzle.

Standard operating procedures for highrise buildings, particularly those not protected by automatic sprinklers, should reflect the potential need to employ heavy firefighting streams, which may require higher flows and pressures.

6. **Pre-fire planning is an essential fire department function.**

The availability of information about the building was a problem in this incident.

The purpose of conducting pre-fire plans is to gather information about buildings and occupancies from the perspective that a fire will eventually occur in the occupancy. This information should be used to evaluate fire department readiness and resource capabilities. At a fire scene, pre-fire plan information can be used to formulate strategies for dealing with the circumstances which present themselves.

Pre-fire planning activities should identify building and fire protection features which are likely to help or hinder firefighting operations and record this information in a format usable to fire-fighters at the scene of an emergency. Recognizing and recording information about restricting devices and pressure reducing valves should be among the highest priorities. Information on fire alarm systems and auxiliary features such as elevator recall, fan control or shutdown, and door releases should also be noted.

The fire department was unable to obtain important details about the installed fire protection at One Meridian Plaza during critical stages of the fire attack. Detailed information about the design and installation of standpipes, pressure relief valves and the fire pump, could have aided firefighters significantly if it had been available earlier in the fire.

Pre-fire plans and standard operating procedures should also consider evacuation procedures and plans for the removal of occupants.

7. **Standard Operating Procedures (SOPs) and training programs for fires in highrise buildings should reflect the installed protection and highrise fire behavior.**

Training and SOPs should consider ways to achieve adequate fire flow with available pressures and ways to improve flow and pressure when required. Tactics for placing multiple lines in service simultaneously must also be developed and discussed.

Extensive pre-fire planning and training are required for fire department control of mechanical smoke management systems to be effective. Training in the management of smoke should consider stack effect and the ability to predict and/or direct ventilation in a real incident.

8. **Safety-oriented strategies should dominate command decisions when multiple systems failures become evident.**

This incident presented command officers with an unprecedented sequence of system failures. As more things went wrong, officers had to seek alternative approaches to manage the situation. The time pressure and stress of fireground command can make it difficult to thoroughly evaluate each alternative approach, particularly as new and unanticipated problems are presented in rapid succession. Conservative tactics, oriented toward protecting the firefighters who must execute them, should take precedence when confronted with an unknown and unanticipated situation, since the potential consequences of fireground decisions can rarely be fully evaluated during the incident. As much as possible, these alternatives should be considered beforehand in pre-fire planning, standard operating procedures, and training materials, and by reviewing post-fire critiques and reports of other incidents. This is an incident that will make a major contribution to the knowledge of what can and will happen when major system failures occur in the worst imaginable sequence.

9. **Fire code enforcement programs require the active participation of the fire department.**

In Philadelphia, responsibility for the fire code is fragmented. The fire department is responsible for developing and maintaining fire code requirements, supervising the appeals process, and investigating and referring fire code complaints. However, it does not conduct regular periodic code enforcement inspections, issue permits, or develop target hazard protocols for ensuring that inspections conducted by the responsible agency are addressing critical fire protection problems.

Many of the model code requirements that apply to highrise buildings are predicated upon assumptions about firefighting strategies and tactics. Most model code organizations designate the fire department, fire prevention bureau, or fire marshal's office as the principal enforcement authority for fire protection system requirements. Fire department personnel are in the best position to validate code assumptions and see that the built-in fire protection and life safety systems are functional and compatible. Moreover, the first-hand knowledge and experience of firefighters with fire behavior is often an invaluable resource when interpreting fire and building code requirements.

**10. Code provisions should be adopted requiring highrise building owners to retain trained personnel to manage fire protection and life safety code compliance and assist fire department personnel during emergencies.**

Model fire prevention codes require building owners to develop highrise fire safety and evacuation plans to manage the life safety complexities of these buildings. The requirements are not specific about what must be included in these plans, and they give no explicit consideration to the problems of firefighting and property conservation.

Mandating the appointment and certification of individuals with specialized knowledge in code requirements and building systems would go a long way toward ensuring that the unique aspects of each highrise building are incorporated into detailed plans.

(New York City Local Law 5 requires that each owner designate a fire safety director. The responsibilities of this individual for managing the life safety plan during an incident are clearly established in this ordinance.)

**11. Occupants and central station operators must always treat automatic fire alarms as though they were actual fires, especially in highrise buildings.**

Building personnel, alarm services, and fire departments must develop an expectation that an automatic alarm may be an indication of an actual fire in progress. Automatic detection systems have gained a reputation for unnecessary alarms in many installations. This has caused an attitude of complacency that can be fatal in responding to such alarms. To change such attitudes and expectations, it will be necessary to improve the reliability and performance of many systems.

*By choosing to investigate and verify the alarm condition, the building engineer nearly lost his life. If not for the ability to communicate with the lobby guard to relay instructions for manually recalling the elevator, this individual would likely have shared the fate of his counterpart who died in a service elevator at the First Interstate Bank Building Fire in Los Angeles (May 4, 1988).*

Technological advances and improved maintenance procedures are the answers to solving the nuisance alarm problem. In addition to requiring regular maintenance of systems by qualified individuals, Philadelphia and other cities have stiffened the penalties on owners, occupants, and central station operators who fail to report automatic fire alarm activations.

**12. Incomplete fire detection can create a false sense of security.**

Automatic fire detectors, like automatic sprinklers, are components of engineered fire protection systems. A little protection is not always better than none. Over-reliance on incomplete protection may lead to a false sense of security on the part of building owners and firefighters alike.

Automatic fire detectors can only notify building occupants or supervisory personnel at a central, remote, or proprietary station that an event has occurred, and in some cases initiate action by other systems to limit the spread of fire, smoke, or both. (In this case, automatic detectors initiated an alarm, recalled elevators, and shutdown air handling equipment; however an elevator was subsequently used to go to the fire floor to investigate the alarm.)

Smoke detectors at One Meridian Plaza were installed in particular areas as required by the 1981 amendments to the fire code; that is at the point of access to exits, at the intakes to return air shafts, and in elevator lobbies and corridors. The apparent underlying logic was to protect the means of egress and to detect smoke in the areas where it was most likely to travel. It appears in this case that the partitions and suspended ceiling contained the smoke and heat during the fire's incipient phase and prevented early detection. In all likelihood, the first detector may not have activated until after the room of origin had flashed-over. Shortly after flashover, the suspended ceiling in this area probably failed permitting the fire to spread throughout the return air plenum. Once the fire broke the exterior windows and established an exterior air supply there was little that could be done to control the fire. Firefighters were disadvantaged by the delay in reporting the fire.

### 13. Nationally recognized elevator code requirements for manual control of elevators during fire emergencies work.

Elevator control modifications at One Meridian Plaza were accomplished in accordance with Commonwealth of Pennsylvania requirements based on ANSI/ASME A17.1, *Safety Code for Elevators and Escalators*. The elevators performed as expected by the standard. The only problem with the elevator response was the decision of the building engineer to override the system to investigate the alarm.

### 14. The ignition source provided by oil-soaked rags is a long recognized hazard that continues to be a problem.

Had the contractor refinishing paneling on the 22nd floor not carelessly left oil soaked cleaning rags unattended and unprotected in a vacant office, this fire would not have occurred. To pinpoint the particular source of ignition of this fire as the sole cause of the death and destruction that followed is a gross oversimplification. Nevertheless, failure to control this known hazard is the proximate cause of this disaster. The danger of spontaneous heating of linseed oil-soaked rag waste is widely recognized. Each model fire prevention code requires precautions to prevent ignition of such materials. At a minimum, waste awaiting removal from the building and proper disposal must be stored in metal containers with tight-fitting, self-closing lids. Leaving these materials unattended in a vacant office over a weekend was an invitation to disaster. This is both an education and an enforcement problem for fire prevention officials.

### 15. Building security personnel should be vigilant for fire safety as well as security threats, especially while construction demolition, alteration, or repair activities are underway.

Earlier in the day, the building engineer had become aware of an unusual odor on the 22nd floor which he associated with the refinishing operations which were underway there. When the alarm system activated later that evening he first believed the solvent vapors had activated a smoke detector.

The roving security guard made no mention of anything unusual during his rounds of the fire area earlier in the evening. It is conceivable that no detectable odor of smoke or audible or visible signals of a fire were present when the guard last checked the floor. However, a cursory check is not adequate when construction, demolition, renovations, or repair activities are underway in a building area. Fire hazards are often associated with construction activities, and buildings are especially vulnerable to fire during such operations. For these reasons, it should be standard practice to check these areas even more carefully and thoroughly than usual. All building operating and security personnel should have basic training in fire prevention and procedures to be followed when a fire occurs.

### 16. Emergency electrical systems must be truly independent or redundant.

Article 700 of the National Electrical Code recognizes separate feeders as a means of supplying emergency power. However, Section 700-12(d) requires these services to be "widely separated electrically and physically...to prevent the possibility of simultaneous interruption of supply." Installing the primary and secondary electrical risers in a common enclosure led to their almost simultaneous failure when the fire penetrated voids in the walls above the ceiling of the 22nd floor electrical closet. The intense heat melted conductor insulation resulting in dead shorts to ground which opened the overcurrent protection on each service interrupting power throughout the building.

Auxiliary emergency power capability was provided by a natural gas powered generator located in the basement mechanical room. This generator was intended to supply one elevator car in each bank, fire pumps, emergency lighting and signs, and the fire alarm system. However, this generator set failed to produce power when needed. (Generator maintenance records indicated a history of problems; however, the root cause or mechanism responsible for these problems was not identified.)

Supplying the generator from the building natural gas service also left the emergency power system vulnerable in the event of simultaneous failure of the electrical and gas public utilities. The transformers that provided power for the adjacent building were installed in the basement of the One Meridian Plaza Building. These transformers had to be shut down due to the accumulation of water in the basement, resulting in the loss of power to this building as well. As a result the elevators in the adjoining building could not be used.

### 17. The regulations governing fire-resistance ratings for highrise structural components should be re-evaluated.

The degree of structural damage produced during the fire at One Meridian Plaza suggests that the requirements for structural fire resistance should be reexamined. Floor assemblies deflected as much as three feet in some places. The fire burning on multiple floors may have produced simultaneous exposure of both sides of these assemblies, which consisted of concrete slabs on corrugated decks, supported by structural steel beam and girder construction, sprayed with cementitious fireproofing materials. The standard fire test for floor and ceiling assemblies involves exposure from a single side only.

Columns and certain other structural elements are normally exposed to fire from all sides. In this fire, the steel columns retained their structural integrity and held their loads. Experience in this and similar highrise fires suggest that columns are the least vulnerable structural members, due to their mass and relatively short height between restraints (floor to floor). Major damage has occurred to horizontal members, without compromising the vertical supports.

The development of uniform criteria for evaluating structural fire endurance accompanied the development of skyscrapers in the early 20th century. Test methods developed at the beginning of the century became the first fire-resistance standard in 1909,[10] which endures today as ASTM E119, *Method of Fire Test of Building Construction and Materials*. One of the principal criticisms of this standard has been its lack of correlation with actual fire conditions.

Many fire protection professionals believe that the standard time- temperature curve used to produce the standard fire exposure during testing is less severe than actual fires involving contemporary fuel loads. (The original test method was based on less volatile, primarily cellulosic, fuels, while modern plastics and hydrocarbon-based furnishings and finishes produce much more dangerous and severe fire exposures.) Others believe that the current test method works well because it provides a good yardstick for comparing the performance of different systems and has been in widespread use for many decades, generating volumes of data on many systems.[11]

## 18. Inspections must be conducted during and after construction to verify that penetrations in fire-resistance rated assemblies are properly protected.

Voids and so-called poke-throughs in horizontal and vertical fire separation assemblies presented ideal avenues of fire spread during the One Meridian Plaza fire. Openings in the partitions enclosing electrical equipment on the floor of origin permitted the fire to reach and damage the electrical risers, plunging firefighters into darkness early in the fire. Voids in stairwell enclosures permitted smoke to spread in stairwells making firefighting operations difficult and exposing upper floors. Smoke and fire also extended via pipe chases, and telephone and electrical closets.

Through-penetration protection has been a continuing concern and has received considerable attention in building and fire codes in recent years. Each of the model building codes now contains provisions requiring protection of poke-throughs in fire-resistance rated assemblies. Moreover, a whole new industry has been developed to fill the technological void in through-penetration protection which developed with the widespread acceptance of plastic pipe and cable.

The absence of fire dampers in mechanical system supply and return ducts at shaft penetrations on each floor is of particular concern. There is evidence of smoke and fire spread through the air handling system. Nationally recognized model building, fire, and mechanical codes have contained requirements for fire dampers in these locations for many years. The installation of smoke detectors in these locations was an ineffective substitute for protecting the integrity of smoke and fire barriers. This fire clearly illustrates that smoke and fire spread through mechanical system plenums, ducts, and shafts is substantial even without the aid of operating fans.[12]

---

[10] Fitzgerald, R. W. (1981), "Structural Integrity During Fire," in *Fire Protection Handbook*, 15th ed., McKinnon, G. P., ed., Quincy, MA: National Fire Protection Association, p. 5-62.

[11] Actual conditions in most fires produce heat release curves similar to the standard exposure up to the point where oxygen, i.e., ventilation, becomes restricted by fire product release, i.e., smoke and heated gases. However, at this point, actual fires usually diminish in size unless ventilation improves or a renewed oxygen supply is established. Once refreshed with a new air supply, most fires will resume growth, reach a peak, and then diminish as the fuel supply is consumed.

[12] HVAC fans were shut down at night and on weekends, and were not operating at the time of the fire.

**19. Features to limit exterior vertical fire spread must be incorporated in the design of highrise buildings.**

Exterior vertical fire spread or autoexposure can be a significant fire protection problem in construction of highrise buildings if interior fire growth is unrestricted. Because of the difficulty with retrofitting exterior features to restrict fire spread, the installation of automatic sprinklers to restrict fire growth is the most simple approach to managing this risk in existing buildings. Exterior features to prevent fire spread must usually be designed and built into new buildings. Many modern (international style) and post-modern building designs present difficult exterior fire spread challenges because of their smooth exterior facades and large glazing areas. Variegated exterior facades and larger noncombustible spandrels significantly reduce exterior fire spread effects by increasing the distance radiant and conductive heat must travel to stress exterior windows and to heat materials inside the windows on floors above the fire.

## CONCLUSION

The ultimate message delivered by this fire is the proof that automatic sprinklers are the most effective and reliable means at our disposal to protect highrise buildings. When all other systems failed, automatic sprinklers were successful in controlling the fire. The Philadelphia Fire Department was confronted with an essentially impossible situation and did a commendable job of managing the incident. The loss of three firefighters is a tragedy that will always be remembered by the Philadelphia Fire Department. Analysis of the situation reveals, however, that the toll could have been much worse, had it not been for the courage, skills, and experience of this department. Several extremely difficult decisions were made under the most severe conditions. This fire will also be remembered for the lessons that it brings with respect to fire protection systems. To work effectively, such systems must be properly designed, installed, and maintained. When those requirements are not satisfied, the results can be devastating, as clearly demonstrated by this incident.

## SOURCES OF INFORMATION

In addition to references to codes and standards cited in the text, news media accounts of the fire, and interviews with officials of the Philadelphia Fire Department, the following resources were used in the preparation of this report:

Alert Bulletin: Pressure Regulating Devices in Standpipe Systems (May 1991). Quincy, MA: National Fire Protection Association.

Butters, T., and T. Elliott (March 1991). "How the Philadelphia FD Handled the Worst Highrise Fire in the City's History," *IAFC On Scene*, March 15, 1991.

Eisner, H., and B. Manning (August 1991). "One Meridian Plaza Fire," *Fire Engineering*, August 1991. pp. 50-70.

Factory Mutual Engineering Corporation (March 1990). Highrise Buildings, Loss Prevention Data Sheet 1-3. Boston, MA: Factory Mutual Engineering Corporation.

Factory Mutual Engineering Corporation (December 1986). Pressure Reducing Valves for Fire Protection Service, Loss Prevention Data Sheet 3-11. Boston, MA: Factory Mutual Engineering Corporation.

Klem, T. J. (May 1991). Preliminary Investigative Report: One Meridian Plaza, Philadelphia, PA, February 23, 1991, Three Fire Fighter Fatalities. Quincy, MA: National Fire Protection Association.

Klem, T. J. (September/October 1991). "Highrise Fire Claims Three Philadelphia Fire Fighters," NFPA Journal, Sept/Oct 1991. pp. 64- 67, 89.

Linville, J. L., ed. (1991). Fire Protection Handbook, 17th ed. Quincy, MA: National Fire Protection Association.

McKinnon, G. P., ed. (1981). Fire Protection Handbook, 15th ed. Quincy, MA: National Fire Protection Association.

One Meridian Plaza: 12-Alarm Highrise Fire (1991), Philadelphia, PA: Philadelphia Fire Films. Videotape.

# APPENDIX A

# Elevation Drawing of the Building and 22nd Floor Plan, Floor of Origin

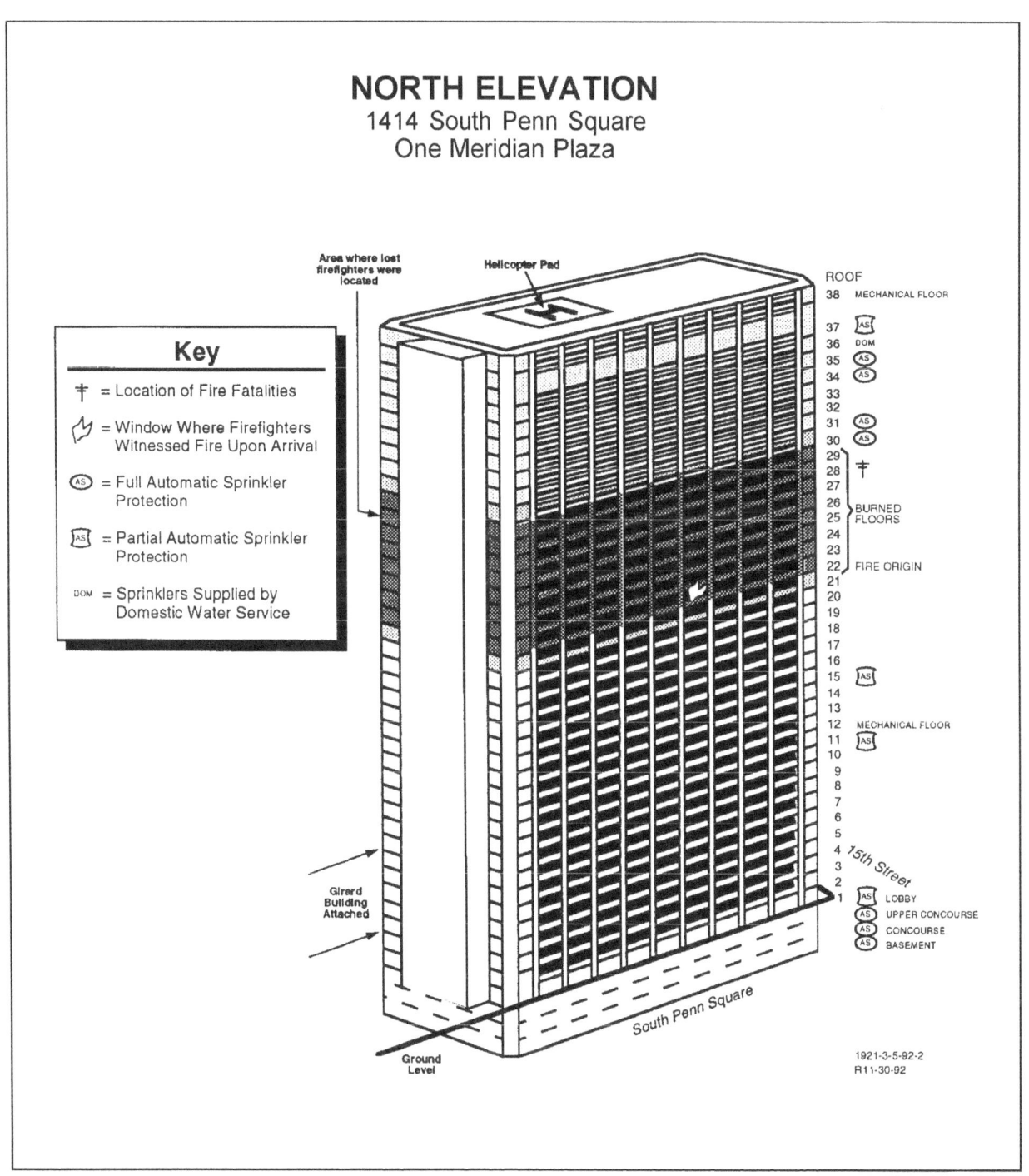

## NORTH ELEVATION
### 1414 South Penn Square
### One Meridian Plaza

**Key**

† = Location of Fire Fatalities

🔥 = Window Where Firefighters Witnessed Fire Upon Arrival

(AS) = Full Automatic Sprinkler Protection

[AS] = Partial Automatic Sprinkler Protection

DOM = Sprinklers Supplied by Domestic Water Service

# Appendix A (continued)

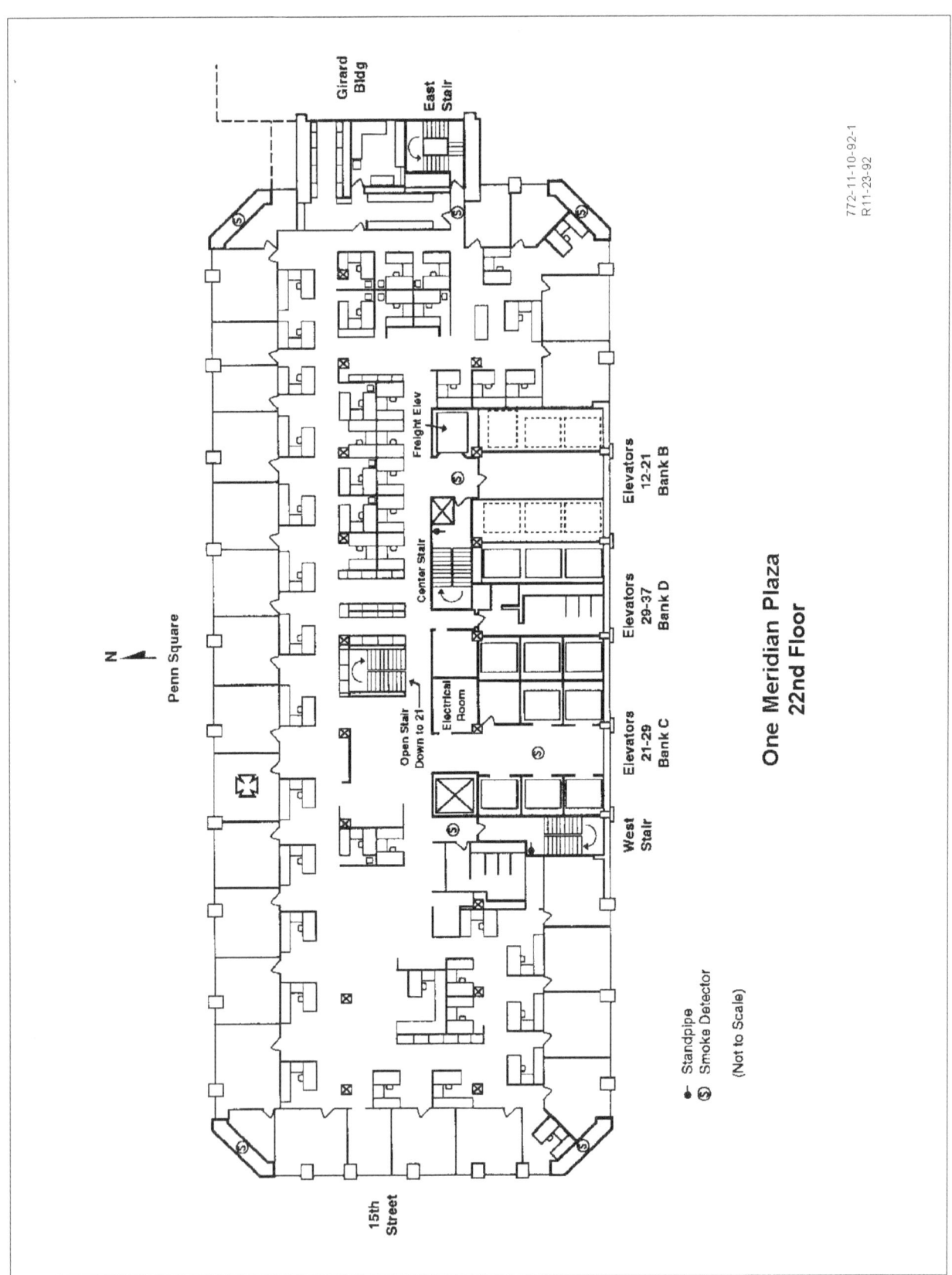

One Meridian Plaza
22nd Floor

Standpipe
Ⓢ Smoke Detector

(Not to Scale)

# APPENDIX B

# Building Emergency Instructions

## FIRE ESCAPE INSTRUCTIONS FOR 3 GIRARD PLAZA
### Fidelity Mutual Life Bldg.

1. If you discover a fire or smell smoke, sound the building alarm. Know the location of the alarm signal stations how they operate.

2. The person at the lobby desk will notify the Fire Department by dialing 9-1-1 when an alarm is transmitted.

3. It is important that the floor captain or alternate floor captain on the *floor* from which the alarm has been sounded notify the person at the lobby desk to report location and nature of the fire. This should be done by going to a safe area, one floor below the fire floor. LOBBY NUMBER IS 585-2365.

4. When fire alarm sounds, Leave at once. Close doors Mind you. Proceed into fire tower and remain there until you are given instructions by the Fire Department or the Building Fire Marshal. Fire towers are safe areas of refuge since they are enclosed and the doors and walls are fire-rated to keep smoke and heat from entering the stairway.

5. DO NOT USE ELEVATORS. They will stop If power fails, causing occupants to become trapped. Elevator shaftways are like chimneys. Smoke could enter the elevator shaft thereby asphyxiating the occupants trying to evacuate the building.

6. Feel the door that leads from your office to the corridor before opening it. If it is hot or smoke is seeping In, do not open. If you become trapped in your office and cannot reach the fire tower, keep door closed and seal off any cracks. Use a phone in the office to call the Fire Department by dialing 9-1-1 and give the address of the building, the floor you are on, and the office number.

7. If door feels cool, open cautiously. Be braced to slam it shut if hall is full of smoke or if you feel heat pressrure against door. If hall is clear, proceed with escape plan.

8. DISABLED PERSONS: A responsible person or persons that work in the same area as the disabled should be assigned to assist in the event of fire. These persons are taken to the fire tower and will remain on the landing.

9. If caught in smoke or heat, stay low where air is better. Take short breaths (through nose) until YOU reach an area Of refuge.

10. AFTER NORMAL WORKING HOURS AND SATURDAYS/SUNDAY'S. All occupants should immediately exit through fire tower doors and proceed directly down and out to street level.

OM 847

APPROVED BY:

---

### DURING NORMAL WORKING HOURS
#### (Mondays through Fridays)

The floor captains and alternates are in complete charge of the evacuation of their respective floors. All personnel will proceed into the fire tower, descending to the next level, stopping one step above the lower floor fire tower door. They are to stand two-abreast so as lo allow room *for* on-coming firemen. Floor captain should be the last one off the floor and will assure the fire tower door Is securely closed.

Remain in fire tower until firemen and or building management give further Instructions.

### AFTER NORMAL WORKING HOURS
### A N D
### SATURDAYS/SUNDAYS

ALL occupants should immediately exit through fire tower doors and **PROCEED DIRECTLY DOWN AND OUT TO STREET LEVEL**

**PLEASE DO NOT USE ELEVATORS.**

ALL new employees should be informed of these procedures.

### FIRE DRILLS

Fire drills are held every two (2) months. These arc pre scheduled AND ALL OCCUPANTS SHOULD BE NOTIFIED OF THE TIME AND DATE.

We arc all to follow the foregoing fire emergency procedures.

**THERE WILL NOT BE AN "ALL CLEAR"** issued by the building. Floor captains are to advise persons in tower to return to floor if floor evacuation is satisfactory.

The date and time of fire drills will always be announced in advance. If you hear a fire alarm and there has been no prior notification, you must assume that there is a real fire emergency.

# APPENDIX C

# Philadelphia Fire Department
# Highrise Emergency Procedures

PHILADELPHIA FIRE DEPARTMENT

OPERATIONAL
PROCEDURE 33
DECEMBER, 1981

SUBJECT: HIGH RISE EMERGENCY PROCEDURES

I. PURPOSE:

To provide guidelines and policy for Fire Department operations at high rise fires and/or emergencies.

II. RESPONSIBILITY:

It will be the responsibility of each member to exercise the appropriate control dictated by his rank in the implementation of this Operational Procedure

III. DEFINITIONS:

A. HIGH RISE BUILDING:

A high rise building is one in which total emergency evacuation is not practical and in which fire must be fought internally because of the building height. The usual characteristics of such a building are:

1. Portions are beyond the reach of Fire Department aerial equipment.

2. Poses a potential for a significant stack effect.

3. Requires unreasonable evacuation time.

B. OPERATIONS COMMAND POST:

An Operations Command Post will be established at the scene of all high rise building emergencies. Conditions permitting, the ideal location is on the ground floor of the building in the lobby at or near the main desk. Typical information available at an Operations Command Post should include floor plans, type of occupancy, names and phone numbers of key personnel, persons presently in the building, unusual conditions and/or circumstances, information on ventilating System, utilities, elevators, etc. This information will be supplied by she building owners. The overall fireground operation will be coordinated from this position in conjunction with the Tactical Command Post.

# Appendix C (continued)

OPERATIONAL
PROCEDURE  33
DECEMBER,  1981

C.  TACTICAL  COMMAND  POST:

The Tactical Command Post will serve as the central
location from which the coordination and tactical
decisions will emanate for combating the emergency
condition.  The Tactical Command Post will be set up
on the floor below the fire or where conditions dictate.
When the Officer in Charge of the Tactical Command Post
makes periodic size-up excursions, he will maintain
communications with the operating forces in his area as
well as the Operations Command Post.

D.  STAGING  AREA:

The staging area will be located in close proximity to
the Tactical Command Post.  First aid station, equipment,
stand-by manpower and logistical support will be marshalled
here.

V.  PROCEDURES:

A.  GENERAL:

1.  Preplanning

High rise operations will be preplanned by the local
company.  Vital Building Information forms will be
updated on an annual basis and station exercises will
be conducted on all platoons to familiarize the members
with conditions and to discuss specific fire fighting
operations and/or situations that may be encountered.
Preplanning tours will be coordinated through the Fire
Commissioner.

2.  Communications

Communications will be maintained at all times between
operating units and the Fire Communications Center (F.C.C.).
All pertinent information will be routed through the
Operations Command Post.  Communication options to keep in
find are vehicle radios, portable radios, Bell Telephone,
elevator phones, intercoms and public address systems.
The spare portable radios on F-100 can be utilized by
operating units to facilitate communications.  Cooperation
and communication with building management and maintenance
personnel is viral in high rise situations.

# Appendix C (continued)

```
                                          OPERATIONAL
                                          PROCEDURE  33
                                          DECEMBER, 1981


    B.   STANDARD   OPERATING   PROCEDURES:

         1.   The first arriving unit will size up the situation
              and give a complete report to the F.C.C.  Members
              are to bear in mind that high rise buildings are
              tightly  constructed and if there is visible evidence
              of fire or smoke, this could indicate a serious fire.
              Anticipate the time required for responding units to
              get into service and-do not hesitate to call for
              additional help.  The Vital Building Information form
              for the involved building will be made available at
              the Operations Command Post at the onset of operations.

         2.   First arriving units, Engine and Ladder, will prepare
              for standpipe and/or sprinkler operations following
              accepted P.F.D. procedures and then proceed to locate
              the fire.   If the Battalion Chief has not arrived at
              this point, one man will be left on the ground floor
              to provide the incoming Chief with all available
              information.

         3.   The first arriving Chief will be responsible for
              designating the location of the Operations Command
              Post.  He will station a man there and proceed to
              the problem area setting up the Tactical Command Post.

         4.   The second arriving Chief will man the Operations
              Command Post upon his arrival.  He will communicate
              with and assist the Battalion Chief at the Tactical
              Command Post wherever possible.  He will coordinate
              the incoming units directing them where needed.   The
              Deputy Chief and/or subsequent arriving command
              personnel or their designated representative will
              assume duties at the Operations Command Post.

         5.   Other first alarm companies as well as subsequent
              alarm companies will, in the absence of specific
              orders via fire radio, proceed in and standby.

    V.  GUIDELINES:

         A. TOOLS AND EQUIPMENT

              1.   Members should not go above the ground floor in a
                   high rise fire without the proper tools and equipment.
                   In addition to required hoseline, forcible entry
                   situations may be met as well as heavy smoke and poor
                   visibility conditions.
```

# Appendix C (continued)

2.  An engine company's high rise tool compliment will consist of the following equipment:

    a.  Standpipe adapters for particular buildings.

    b.  Three (3) rolled lengths of 1-3/4" hose with shut off.

    c.  A gated-wye or half of- the reducing adapter.

    d.  Roof rope.

B.  STAIRWAYS:

1.  Every effort should be made to maintain the integrity of stairways and towers, as these are main evacuation routes. Doors leading into these exitways should not be propped open, as the introduction of smoke and heat into these avenues of life safety might preclude their use.

    In the event a stairway or tower is to be used, considaration must be given to prior evecuation of the upper floors, where required, or the availability of another means of egress, remote from the area of involvement.

2.  Towers are good locations to initiate fire attack as standpipes are usually located there and an escape route is readily available.

C.  ELEVATORS:

1.  The location and status of all elevators should be determined early in the operation, because of the possibility of people being trapped in stalled elevators. Every effort should be made to return all elevators to the ground floor so they can be controlled by Fire Department personnel.

2.  The manual over-ride key will be obtained at the lobby console or from the building engineer. Freight elevators are best suited for P.F.D. operations, in that, they usually serve *all* floors and have greater carrying capacity. Where possible, elevator banks remote from the fire should be utilized.

# Appendix C (continued)

OPERATIONAL
PROCEDURE  33
DECEMBER, 1981

3.  If there is any doubt about the safe use of elevators, members will utilize stairways and towers.

D.  OPERATIONAL  CONSIDERATIONS:

1.  Ventilation

a.  Building personnel may be able to indicate what internal mechanical ventilation can be affected. Frequently, ventilating systems can be reversed to exhaust smoke.  If details of the system cannot be determined, air conditioning and ventilating systems should be shut down to curtail the spread of smoke and heated gases throughout the building.

b.  Smoke ejectors properly utilized can be of great value in channeling smoke.  Window air conditioners may be of value in the exhaust position.

c.  Breaking glass on the upper floors of a high rise building is an extremely risky practice.  Even if police lines are maintained, glass falling from extreme heights could carry over long distances causing serious injury to both civilian and Fire Department personnel.  Glass should be broken only as a last resort.

VI.  EVACUATION.

A.  RESPONSIBILITY:

The building owner or manager will have the responsibility of preparing an evacuation plan for the high rise building occupants and/or tenants, with the assistance of the Fire Prevention Division.  The high rise building Fire Marshal will have the initial responsibility of occupant movement in a fire emergency and may order total evacuation, if conditions warrant, before the arrival of the P.F.D.

B.  PROCEDURES  TO  FOLLOW:

1.  In buildings with two or more fire towers, all building occupants will enter the fire tower and line the stairway at the sounding of the building fire alarm, prepared to evacuate.  When the location of the fire is confirmed, P.F.D. personnel will institute the removal and relocation of building occupants in these fire towers.  Occupants will be assisted to refuge areas below the fire floor.

# Appendix C (continued)

OPERATIONAL
PROCEDURE  33
DECEMBER,  1981

a.   When the occupant load is such that the fire towers
     will not accommodate all of the occupants, the
     initial evacuation into the fire towers will be the
     fire floor and two floors above the fire floor,
     before the arrival of the P.F.D.

2.   In buildings with one interior enclosed fire tower or
     buildings that have only open stairways, total
     evacuation will be started by the Building Fire Marshal.
     Upon arrival, the Commanding Officer will determine the
     need for continued and/or total evacuation.

3.   In buildings of newer construction that possess the
     capability of pressurizing individual floors and ele-
     vator shafts, the building Fire Marshal will institute
     a three floor evacuation plan.  The building fire alarm,
     when activated, will sound on the fire floor, the floor
     above the fire floor and the floor below.  The occupants
     on the affected floors will proceed into the fire towers
     and line the stairway.  A warning tone will sound on the
     remaining floors when the alarm is activated and the
     occupants on these floors will stand by their-assigned
     fire tower for further instructions, via the public
     address system, by the building Fire Marshal or the
     Fire Officer in charge.  At the sounding of the alarm,
     the building engineer will shut down the air handling
     system on the fire floor and pressurize the floor above
     the fire floor and elevator shafts.

4.   In all high rise hotels and motels, the owner or manager
     will supply written evacuation procedures for the hotel
     guests which will be posted on the inside of each guest
     room door, and the elevator lobby at each level.  When
     hotel guests are taken to a refuge area, at least one
     member from the P.F.D. will remain with the guests and
     inform them when they may return to their rooms.

5.   The first arriving fire officer will obtain the list of
     all disabled persons at the lobby console or from the
     building Fire Marshal.  It will be necessary for P.F.D.
     members to assist the disabled persons from the fire
     tower to the ground floor, if partial or total evacuation
     is necessary.  This can be accomplished by using elevators
     remote from the fire area.  This procedure will include
     all high rise hotels.

     REFERENCE:   High Rise Fire Safety Training Manual
     Philadelphia Fire Code - Chapter 5-3400 High Rise Building.

          BY ORDER OF THE FIRE COMMISSIONER

# APPENDIX D

# Factory Mutual Engineering Corporation
# Pressure Reducing Valve Loss Prevention Data Sheet

## Loss Prevention Data

### PRESSURE REDUCING VALVES
### FOR FIRE PROTECTION SERVICE

#### SCOPE

This data sheet provides guidelines for the installation, maintenance and testing of pressure reducing valves for fire protection service. The intent here is not to encourage installation of pressure reducing valves, but rather to provide guidelines when their use is unavoidable.

#### GENERAL

1. Uses. Pressure reducing valves are used to reduce high water pressure at their inlet side to a lower acceptable pressure at their outlet side. Pressure reducing valves are used on standpipes and sprinkler systems in high-rise buildings, below grade mines, and are also used to regulate pressures in underground piping. Other uses may exist if warranted by specific local conditions.

Data Sheet 4-4N, Standpipe and Hose Systems, limits the maximum pressure at the standpipe hose outlet to 100 psi (689 kPa) (6.9 bar). The majority of approved sprinkler system components are rated at 175 psi (1206 kPa) (12.1 bar) working pressure. Pressure reducing valves may be installed on hose inlets, on sprinkler system inlets, on common feed to hose connections and sprinkler systems, or in piping downstream of a fire pump or high pressure water system. In any case, they are set to limit the maximum pressures to those recommended for the particular fire protection system involved.

2. Designing to Minimize Use of Pressure Reducing Valves. Proper design of fire protection systems and selection of fire pumps can often eliminate excessively high pressures. In many cases, the need for pressure reducing valves can be eliminated, or the total number needed can be reduced.

In high-rise buildings with combined standpipe and sprinkler system risers, it may be necessary to install pressure reducing valves to limit maximum available pressure at the hose connection to 100 psi (6.7 bar) (670 kPa). On the other hand, sprinkler system connections have no such pressure restrictions other than the equipment ratings, which are usually 175 psi (12.1 bar) (1210 kPa). Thus, pressure reducing valves would not be needed on sprinkler system connections at any floor where the static pressure is 175 psi (12.1 bar) (1210 kPa) or less.

3. Types. There are two basic types of pressure reducing valves: direct-acting and pilot-operated. Direct-acting type valves have internal controls consisting of a spring or other type of mechanical device that acts directly on a piston or diaphragm to restrict waterflow through the valve, thus controlling pressure at the valve outlet.

Pilot-operated valves operate hydraulically using the **outlet** pressure to control the position of an internal waterway disc to restrict waterflow through the valve, thus controlling pressure at the valve outlet.

4. Characteristics

(a.) Direct-acting valves. Direct-acting pressure reducing valves are generally angle-type valves. A schematic of a type of direct-acting pressure reducing valve is shown in Figure 1. This valve is typical of most direct-acting pressure reducing valves, in that it acts both as a

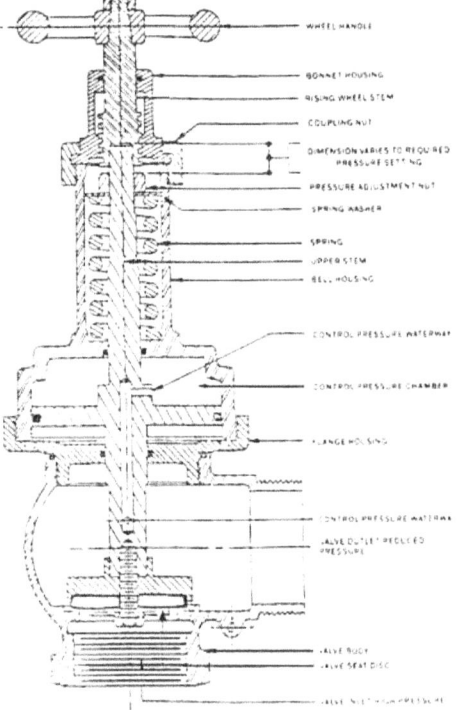

Figure 1. Direct-acting type pressure reducing valve.

# Appendix D (continued)

gauge readings whenever water supply control valves are operated, and bleed off excess pressure downstream of the pressure reading. Regular Inspection, testing and maintenance will help detect any problems so that immediate remedies may be accomplished.

## ILLUSTRATIVE LOSS

Erratic Operation of Pressure Reducing Valve Temporarily impairs Water Supply.

A shower of sparks from an overhead crane railignited oil-coated titanium turnings inside this scrap metal recycling facility. An intense fire spread to the combustible roof and roof covering due to the temporary impairment of water supplies to operating sprinklers caused by erratic operation of the pressure reducing valve installed in the supply main from the public water system. Many sections of the wood plank roof were in deteriorated condition, some to the point where planks had fallen out, and the roof covering could be seen from below. Combustible roof and roof coverings burned unchecked over an 80,000-ft² (7440-m²) area until the fire department could bring the blaze under control. The pressure reducing valve was later found to have a faulty seal, which caused the erratic operation of the valve. Property damage S3,500,000 (01751X-81-1-8)

## RECOMMENDATIONS

### General

1. Whenever possible, design water supply and fire protection systems to avoid the need for pressure reducing valves.

2. For fire protection service, pilot-operated pressure reducing valves are recommended to take full advantage of the water supply available at the valve setting. Direct-acting pressure reducing valves, by the nature of their design, will cause the pressure to be reduced below that of the valve setting. The amount of this excess pressure reduction is due to friction loss and will increase with increasing flow. This drawback complicates system design by not taking full advantage of the available water supply.

3. When a pressure reducing valve is to be installed, determine the characteristics of the inlet supply and calculate the maximum water demand. The valve setting for a pressure reducing valve is fixed by two limits: (1) the maximum pressure to be permitted on the system downstream from the pressure reducing valve; and (2) the residual pressure required on the outlet side of the pressure reducing valve at the maximum water demand flow rate. These limits, together with information regarding the water supply available on the inlet side of the pressure reducing valve, and the hydraulic and friction loss characeristics of the particular pressure reducing valve, should be evaluated so that a

pressure reducing valve of adequate capacity and suitable type for the specific installation conditions can be provided.

4. Provide each pressure reducing valve with a permanently attached placard that indicates valve setting pressure.

### Installation

Pressure reducing valves can be installed either in pits on underground piping systems, or on individual sprinkler systems, such as would be typical at each floor of a high-rise building. Specific installation guidelines that apply to each type of installation follow.

### UNDERGROUND PIPING SYSTEMS

1. When a pressure reducing valve is needed to reduce only non-fire service water pressures, provide a separate fire service water connection without a pressure reducing valve.

2. When a pressure reducing valve is needed for both non-fire service and fire service water, arrange the pressure reducing valve and non-fire service water connection as shown in Figure 3. This will allow isolation of the non-fire service water connection without affecting fire service water, this also allows regular exercising of the pressure reducing valve through normal draft.

3. Provide a bypass loop around the pressure reducing valves, with a normally closed indicating control valve, to allow water for fire protection to be available in the event the pressure reducing valves are out of service.

4. Large pressure reducing valves may not provide accurate pressure regulation or may cavitate at low waterflow rates. Thus, it is necessary to provide a smaller pressure reducing valve in parallel with the primary pressure reducing valve, as indicated in Figure 3. The smaller pressure reducing valve should be capable of regulating the pressure in the range of no flow up to the maximum flow for which it is recommended. The larger pressure reducing valve should be capable of regulating the pressure in the range of the upper limit of the smaller pressure reducing valve up to flow for the maximum water demand. Each installation should be carefully engineered, taking into consideration the water supply, the water demand and the characteristics of the particular pressure reducing valves. If necessary, consult with the valve manufacturer to ensure correct valve settings and installations.

5. Provide indicating control valves as shown in Figure 3 to allow

(a.) Isolation of each pressure reducing valve for maintenance.

# Appendix D (continued)

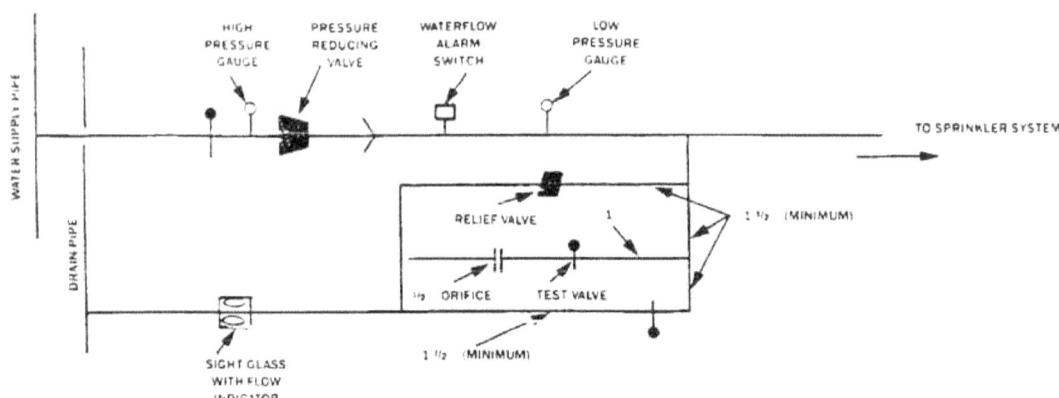

*Figure 4. Arrangement of pressure reducing valve for individual sprinkler systems, typically seen on each floor of a high-rise building.*

The test valve and associated piping, should be 1 in. (25 mm) in size with a I/2 in. (13 mm) restricted orifice downstream. This will allow testing of the waterflow alarm by simulating the flow through one sprinkler.

The drain valve and associated piping should be 1½ in. (38 mm) minimum in size, but at least one-half the size of the pressure reducing valve to allow for operational testing of the pressure reducing valve.

5. Locate all pressure reducing valves in dry, accessible areas, arranged for convenient maintenance and testing.

### Maintenance and Testing

1. Proper installation and regular testing of pressure reducing valves are necessary to maintain the valves in good operating condition. In addition to the usual inspections to ensure that water control valves are open and water-flow alarms are functional, inspect pressure reducing valves weekly by opening the test drain. Discharge through the test drain should cause the main valve piston or diaphragm to move, with the degree of pressure reduction indicated by the gauge readings. The test drain should be opened for a long enough time to allow the pressure to stabilize to the valve setting. Reclose the test valve slowly to avoid trapping high pressures downstream from the pressure reducing valve. The manufacturer's instructions should be followed faithfully.

2. Whenever annual water tests are conducted on underground fire protection systems equipped with pressure reducing valves, flow at least the maximum water demand to evaluate the performance of the pressure reducing valve. Ensure that the residual pressure achieved downstream from the pressure reducing valve IS adequate for the maximum water demand.

3. Examine packing glands or stuffing boxes for leaks, but do not tighten them to a degree that would cause sluggish operation of the valve.

4. Discharge through the pressure relief valve indicates a problem which should be repaired promptly.

Note: There IS no NFPA standard on this subject.

FM Engr. Comm September 1986

# APPENDIX E

# Philadelphia Fire Department
# Pressure Regulating Device Fact Sheet

Pressure Reducing Valves

Purpose of Pressure Reducing Valves

Pressure reducing valves (PRVs) are used to reduce high water pressure at the inlet side of the valve to a lower acceptable pressure at the outlet side of the valve.

Where PRVs are Used

Pressure reducing valves are used in water supply systems such as water treatment plants and water distribution systems, including large water reservoir tanks to maintain a constant system pressure by controlling fluctuations in the pressure (figure 1). In sprinkler and standpipe systems their purpose is to control the pressure in the cross mains and branch lines in a sprinkler system and in standpipe systems to prevent excessive hose outlet pressures (figures 2a and 2b).

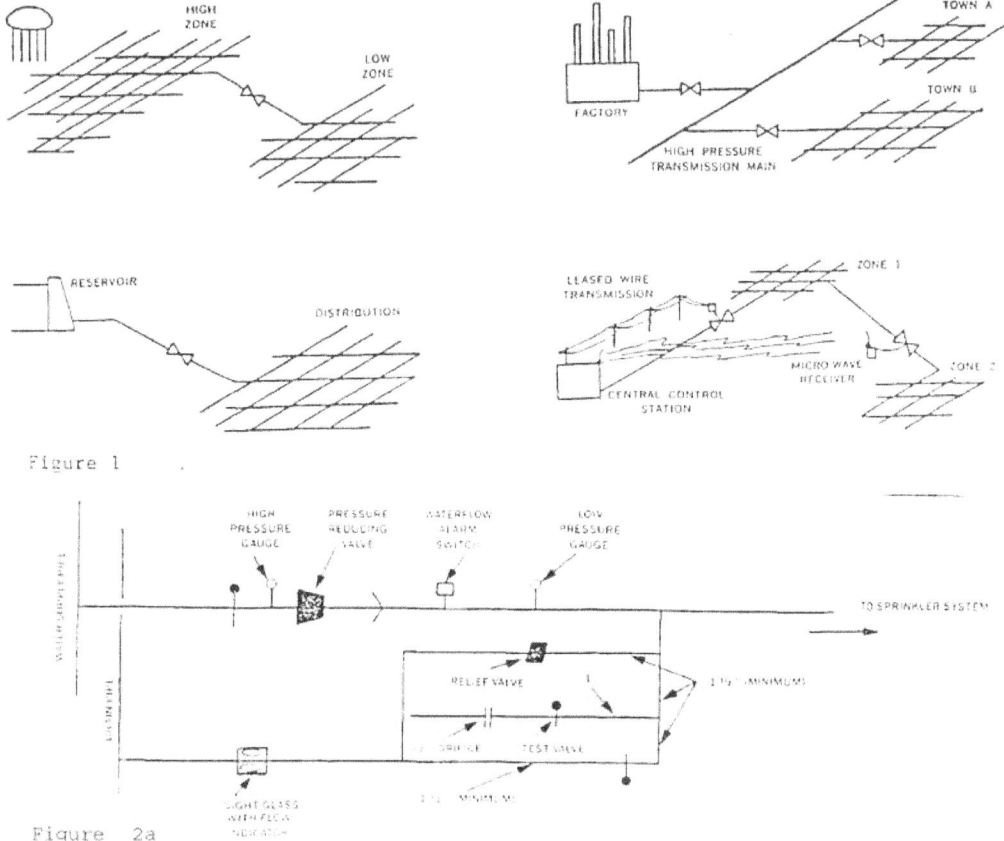

Figure 1

Figure 2a

*Arrangement of pressure reducing valve for individual sprinkler systems typically seen on each floor of a high-rise building*

# Appendix E (continued)

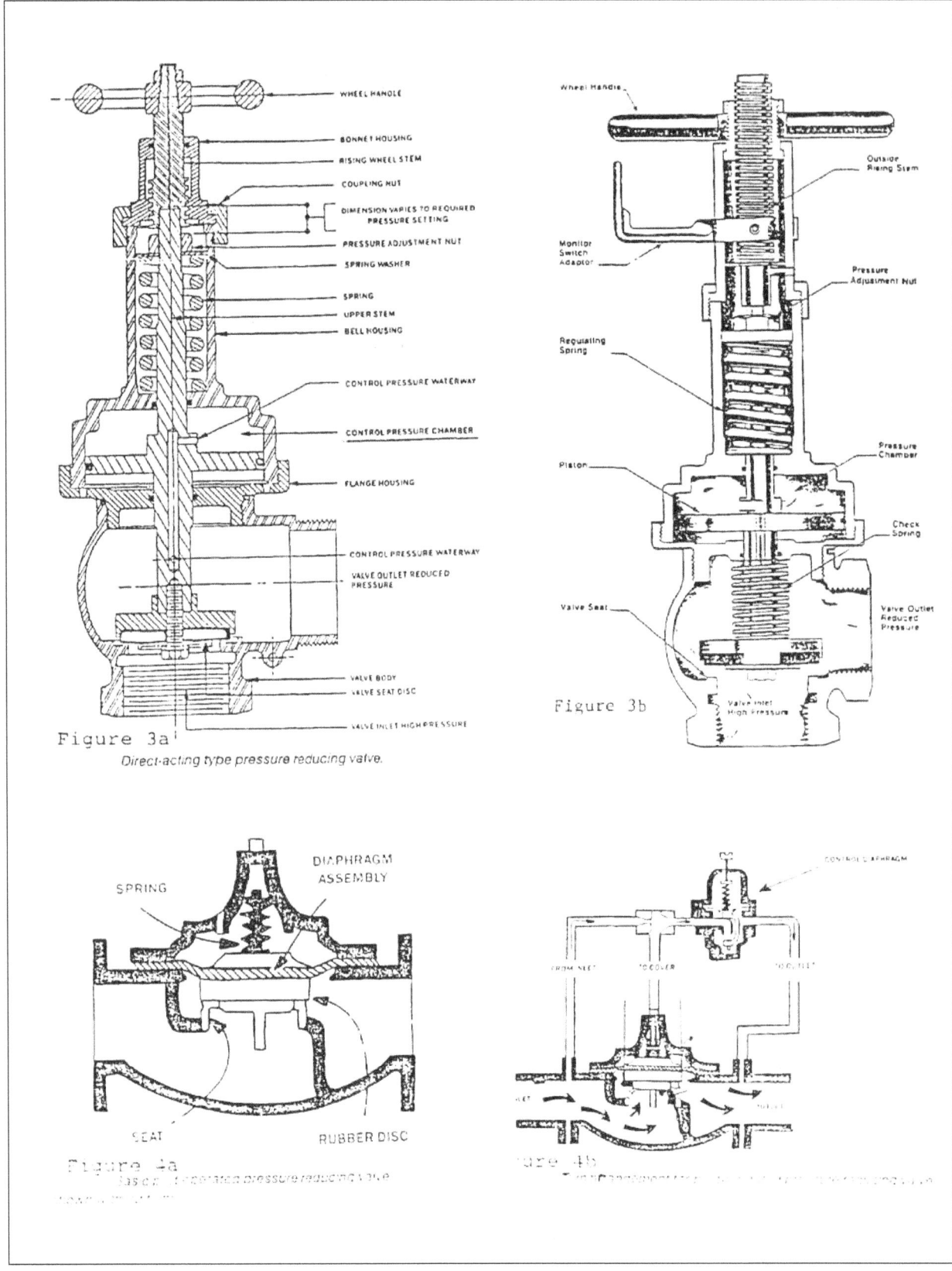

Figure 3a
Direct-acting type pressure reducing valve.

Figure 3b

# Appendix E (continued)

Types of PRVs Commonly Found on Standpipe Systems

The majority of PRVs installed in high-rise buildings in Philadelphia are the direct-acting type. There are two varieties of these devices: pressure restricting and pressure regulating.

Pressure Restricting Devices

Pressure restricting devices are simple devices or valves which control residual pressure by restricting the opening through which the water flows. They do not control (reduce) static pressure.

Examples:   Devices - fixed orifice disk or adjustable orifice valves without shut-off device.
            Valves - adjustable valves with shut-off device.

    Settings

        Orifice disks - non-adjustable - ordered from manufacturer based on a specific pressure and flow.

        Orifice valves - adjustable orifice plate inside valve.

        Adjustable valves - exterior adjustment scale controlling interior valve seat. The adjustment scale usually has a lock or seal to prevent tampering, but these can be overridden.

Pressure Regulating Valves

Pressure regulating valves control residual and static pressures by means of an internal spring and pressure control chamber. The two parts, acting together, regulate high inlet pressure to a lower, acceptable outlet pressure. Some valves also provide a checking feature using an additional, light-weight spring to prevent any back-flow or loss of water in the down stream portion of the system.

    Settings

    Most pressure regulating valves for standpipes are factory-set based on the calculated inlet pressure, desired outlet pressure, flow, and floor to be installed. They are ordered for a specific system and floor, with a label attached to each valve indicating the floor of installation and the calculated outlet pressure at that floor. One brand is field adjustable and is set on site by means of an adjusting rod.

Code Requirements for PRVs on Standpipe and Sprinkler Systems

NFPA

    Standpipe Systems (NFPA 14-1990)

    Where flows at a hose outlet exceed 100 psi, a pressure restricting or regulating valve shall be installed.
    Where flows at a hose outlet exceed 175 psi, a pressure regulating valve be installed.

-4-

# Appendix E (continued)

Sprinkler Systems (NFPA 13-1989)

Pressure reducing valves are required on sprinkler systems where all components are not listed for pressures greater than 175 psi and the potential exists for normal water pressure in excess of 175 psi.  The valve shall be set for an outlet pressure not exceeding 165 psi at the maximum inlet pressure.

Building Code (BOCA-1990)

Standpipe Systems

Where flows at a hose outlet exceed 100 psi, a pressure regulating valve shall be installed.

Exception: Where fire hose is not required at an outlet, a PRV is not required unless the pressure exceeds 175 psi static or residual pressure.

Sprinkler Systems

There are no references in the BOCA Building Code to PRVs for sprinkler systems.

1990 BOCA Building Code Commentary

(The commentary is a companion book to the code which explains various sections of the Code.)

Pressure Regulating Valves on Standpipe Systems.

The pressure at the hose outlet must be regulated in cases where, either due to static head or excessive residual pressure, the pressure at the outlet is more than 100 psi.  The hydraulic calculations should indicate the pressures within the system.  If pressures at hose outlets exceed 100 psi then pressure-regulating devices are to be installed.  Pressure-regulating devices are required to regulate the pressure in both static and flow conditions on each outlet.

The preferred practice in the design of standpipe systems for tall buildings is to divide the system into pressure zones.  Each zone is limited to approximately 12 stories.  Therefore, the water pressure in each zone is not excessive so the need for pressure-regulating devices is eliminated.

Consideration should be given to ensuring that the pressure-regulating device allows the fire department to have full pressure when required.  Fire departments usually require 100 psi at the nozzle and, therefore, a residual pressure of 100 psi at the hose outlet would not be adequate.  Adequate pressure at the hose outlet would be dependent on the length and diameter of the hose and the amount of water flowing.  Generally, at least 150 psi would be required for a nozzle pressure of 100 psi.  For this reason, pressure-regulating devices are not required if occupant use hose is not provided and the static and residual pressures do not exceed 175 psi at the outlet.

U.L. (UL 1468-1985)

Standpipe systems:  A pressure reducing valve shall operate within + 15 psi of the setting pressure of the valve.

Sprinkler systems:  A pressure reducing valve shall operate within + 10 percent of the pressure setting of the valve.

# Appendix E (continued)

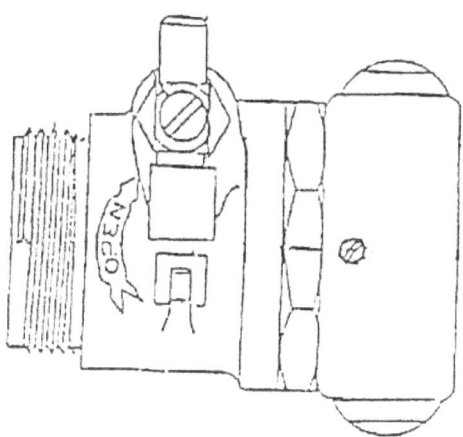 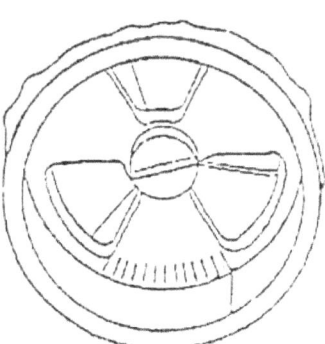

PRESSURE RESTRICTING DEVICE - separate fitting which attaches to the outlet of a standpipe valve.  Device is easily removed to obtain full unrestricted flow.

-6-

# Appendix E (continued)

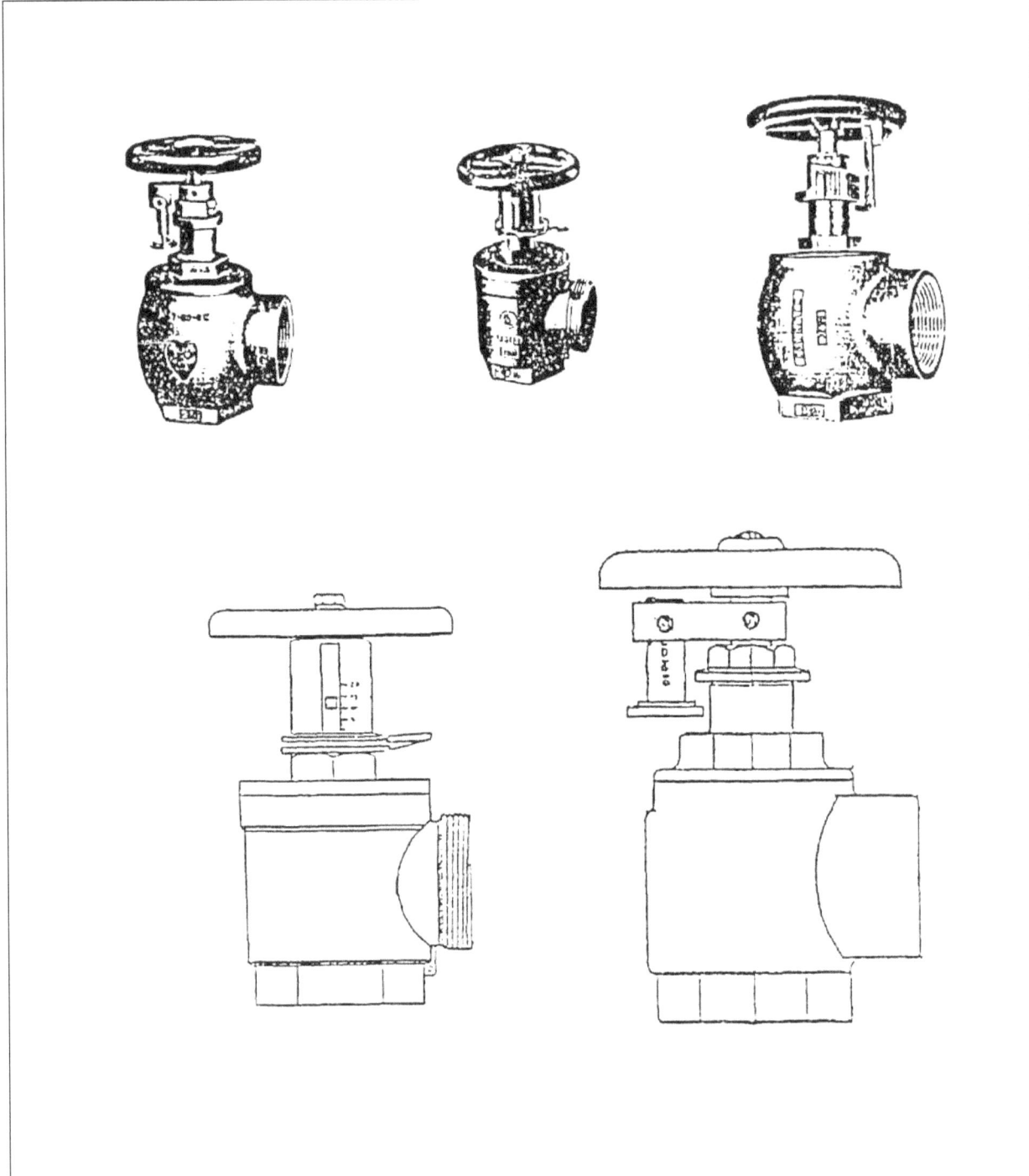

PRESSURE RESTRICTING VALVES - Pressure restricting valves are easily recognizable due to the external fittings on the bonnetts and stems (release links and clips, collars, etc.).

- 7 -

# Appendix E (continued)

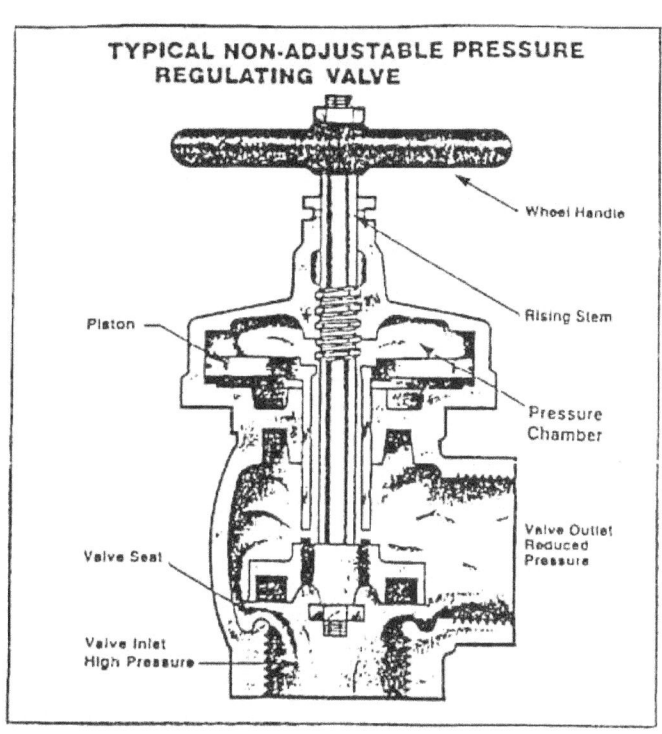

TYPICAL NON-ADJUSTABLE PRESSURE
REGULATING VALVE

Wheel Handle

Rising Stem

Piston

Pressure
Chamber

Valve Outlet
Reduced
Pressure

Valve Seat

Valve Inlet
High Pressure

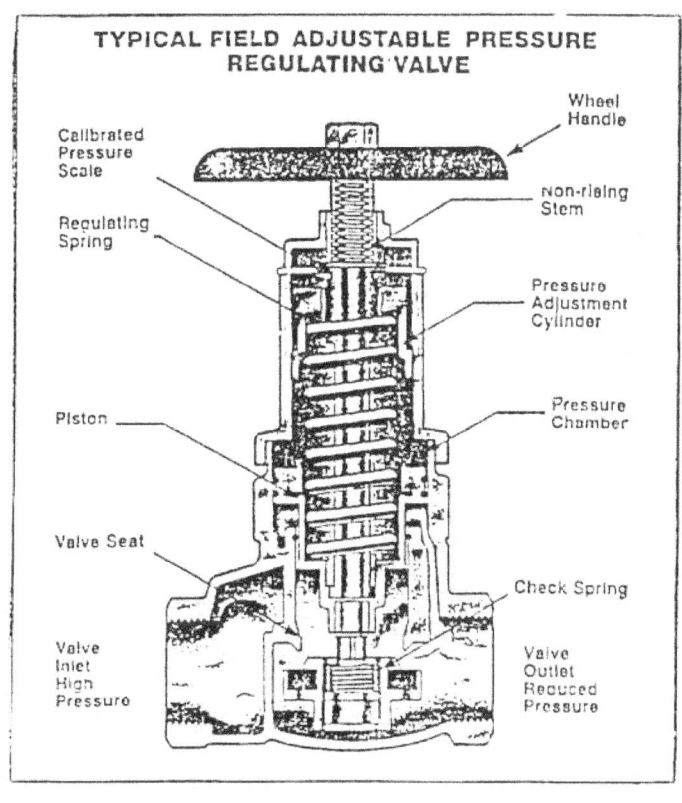

TYPICAL FIELD ADJUSTABLE PRESSURE
REGULATING VALVE

Wheel
Handle

Calibrated
Pressure
Scale

Non-rising
Stem

Regulating
Spring

Pressure
Adjustment
Cylinder

Pressure
Chamber

Piston

Valve Seat

Check Spring

Valve
Inlet
High
Pressure

Valve
Outlet
Reduced
Pressure

# Appendix E (continued)

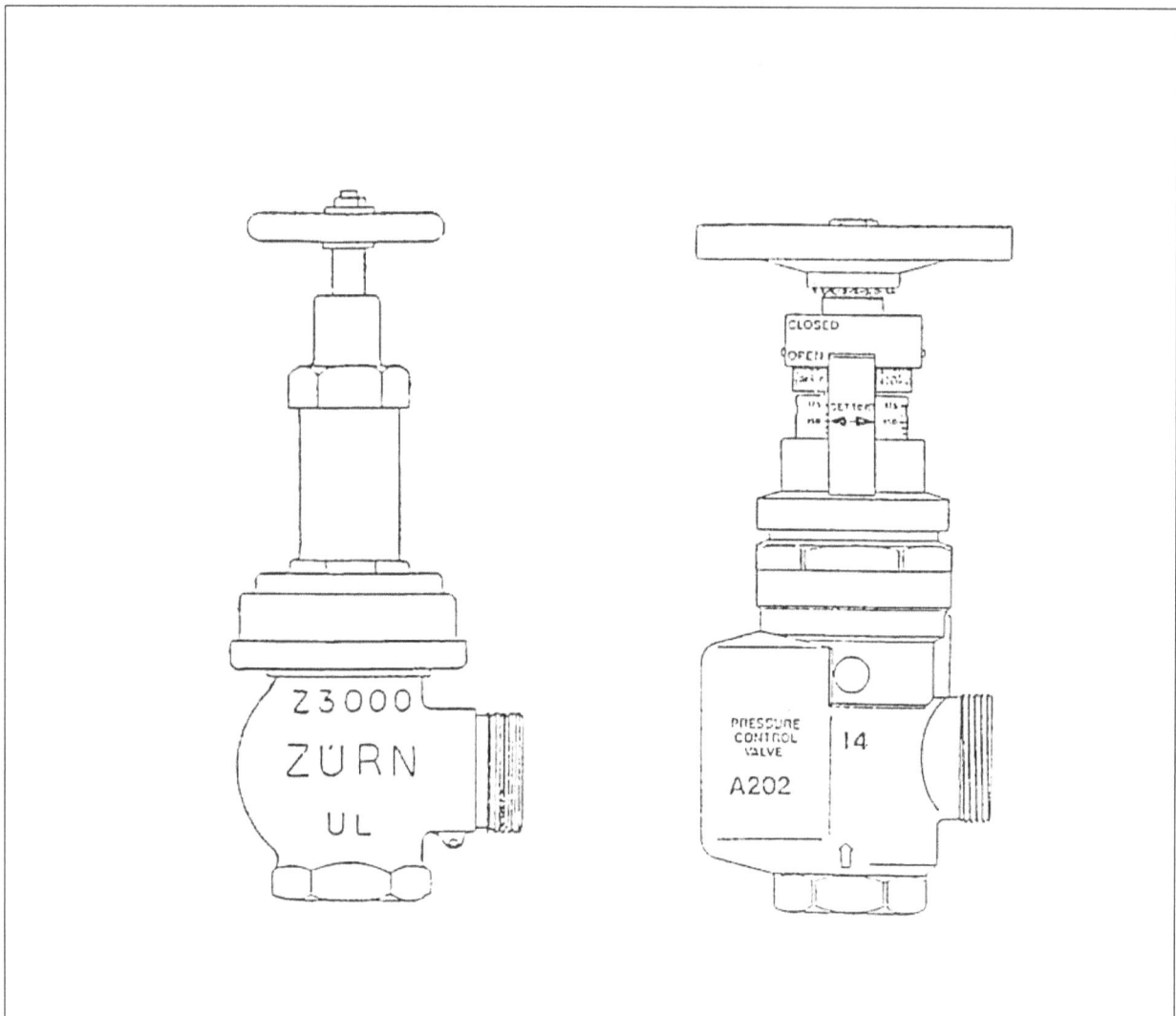

PRESSURE REGULATING VALVES - Direct acting pressure reducing (pressure regulating) valves are required to be permanently marked with: the name or trademark of the manufacturer or private labeler, and a distinctive model number, catalog designation, identification mark, or the equivalent. If unsure whether a particular valve is a PRV, contact the Inspections unit with the above information.

- 9 -

# APPENDIX F

# Philadelphia Fire Department Block Inspection Program

PHILADELPHIA FIRE DEPARTMENT
OPERATIONAL
PROCEDURE 4
AUGUST, 1990

SUBJECT:    BLOCK INSPECTION PROCEDURE

I.    POLICY

Block inspection is an integral part of the Philadelphia Fire Department's Fire Prevention program. The program provides us with additional exposure in the community, enhancing public relations while discovering hazardous conditions that would cause a fire or present a life hazard. Professional appearance, courtesy and concern will have a lasting positive image of Philadelphia Fire Department members.

II.   RESPONSIBILITY

It is the responsibility of each member to exercise the appropriate control dictated by their rank in the implementation of this Operational Procedure.

III.  DEFINITIONS

A.    INSPECTIONS.

For the purpose of record keeping, inspections will be recorded as either regular or referral inspections.

1.    Regular Inspections

Those inspections which reveal no violations or hazards, or minor violations or hazards which can be corrected immediately.

2.    Referral Inspections

Those inspections which reveal serious violations of the Fire Code and/or fire, electrical or building hazards dangerous to life and property.

47

# Appendix F (continued)

B.    BUILDINGS  UNDER  CONSTRUCTION

   1.    Buildings  under  construction  or  renovation  will  be toured  on  a  weekly  basis  by  local  companies.  The Station  Manager  will  determine  which  day  of  the  week the  tour  will  be  conducted.    Center  city  fire companies,  because  of  the  large  number  of  properties under  construction,  will  tour  buildings  being  constructed  or  renovated  as  frequently  as  is  feasible.

   2.    The  purpose  of  the  tour  of  a  building  under  construction  or  renovation  is  to  familiarize  company  members with  the  property.    They  should  note  fire  protection and  exiting  features,  hazards  or  limitations  due  to the  construction,  water  supply  and  any  fire  or  building  code  violations.

   3.    The  standpipe  system  should  be  inspected  to  ensure  it is  operational.    Most  fire  protection  systems  are  not required  to  be  operational  until  a  building  is  issued a  Certificate  of  Occupancy  (full  building  use)  or  a Temporary  Certificate  of  Occupancy  (selected  floor use).    Standpipe  systems  are  required  to  be  operational  when  construction  reaches  the  fifth  floor  or 65  feet  in  height  (BOCA  10,13).    From  that  time  on, the  system  must  be  operational  with  a  Fire  Department intake  connection  and  outlets  on  each  floor  up  to  and including  the  floor  below  the  highest  floor  capable of  being  occupied  (stairs  and  floors  in  place).

   4.    Fire  and  building  code  violations  noted   during  the tour  should  be  handled  in  the  same  manner   as  those found  during  block  inspections.

IV.   PROCEDURES

   A.    INSPECTION  SCHEDULES

      1.    The  Block  Inspection  Program  will  begin  the  second Monday  in  January  each  year.

      2.    Block  inspection  will  be  performed  Monday  through Friday,  excluding  holidays,  at  the  following  times:

      Division  1  -  0950  hours  to  1150  hours  and
                         1250  hours  to  1550  hours.

      Division  2  -  1000  hours  to  1200  hours  and
                         1300  hours  to  1600  hours.

# Appendix F (continued)

3.  The FCC will announce weather conditions only, at 0805 hours and 1205 hours. Block inspections will not be performed when this report indicates temperatures below 40° F. or above 85° F.

4.  Block checks will not be performed during extra alarm fires. Companies will return to their stations as soon as a second alarm is struck.

5.  If, for any reason, a company is not going out on block inspection as scheduled, they will notify the Battalion Chief. Upon request of the Battalion Chief, the company will forward a memo to him.

B.  COORDINATION

1.  Deputy Chiefs will be responsible for the overall coordination within their respective divisions.

2.  Battalion Chiefs are responsible for the coordination and quality of the Block Inspection Program within their battalions on their platoons.

3.  Company officers will insure that all members under their command have a working knowledge of the Block Inspection Program, the Fire and Building Codes and related reports.

C.  GUIDELINES

1.  All buildings in the city will be inspected, annually, with the exception of one and two family dwellings which will be inspected upon request.

2.  Each company's local area will be divided into four sections; one section assigned to each platoon. These sections will be rotated annually so that all platoons inspect their entire district on a four year basis.

3.  Inspectors will be arranged so that half of the companies in the battalion perform inspections in the morning and the remainder in the afternoon.

# Appendix F (continued)

4.   Company inspection times will be alternated weekly, that is, in the. morning one week, in the afternoon the next.

5.   Inspections will normally be performed by the entire company, including the officer, working in teams of two or three. Example: Quota 1 and 3, inspection team will be officer and two firefighters, driver stays with the apparatus. If the quota is 1 and 4, inspection team will be officer and 1 firefighter; and the second team will be two firefighters, driver stays with the apparatus. In those areas where apparatus parking is a problem or the size of a building makes it impossible for inspecting members inside the building to hear the siren, the Battalion Chief will have those companies send individual members out to perform inspections. This will be coordinated through the Deputy Chief. When additional members are detailed to a company for this purpose, the assigned members, rather than the detailed members, will perform the inspections.

6.   The Battalion Chief will order and maintain a sufficient supply of Fire Prevention Check Forms for the battalion. Companies needing forms will request them from their Battalion Chief.

D.   BLOK INSPECTION DUTIES

1.   A Fire Prevention Check Form will be prepared for each building inspected. Where required, the Vital Building Information and Emergency Guide to Hazardous Materials Storage Forms will be prepared and/or updated. This information will be reviewed, as part of the station exercise, by all platoons to familiarize themselves with existing hazards and conditions in their local district.

2.   Unsafe conditions, such as defective traffic signals, missing directional signs, missing sewer inlet covers, missing hydrant caps,. potholes, etc., will be reported. Notify the FCC via the fire radio if a serious hazard exists, other conditions will be reported on a Complaint Report by City Employee Form upon return to station.

# Appendix F (continued)

OPERATIONAL
PROCEDURE 4
AUGUST, 1990

3. Companies encountering buildings without a visible street address will notify the owner that it is in their best interest to clearly identify all buildings to ensure that emergency services are not delayed.

4. Block inspections provide an excellent opportunity for driver training. Company officers will take advantage of the time spent on the street to train new drivers and provide refresher training for qualified drivers. It also enables company personnel to become familiar with their local district.

5. Companies will remain attentive to new construction within their local district and immediately inform Fire Survey of same. If possible, they will secure the information in regard to the numbering system being used and pass this information on to Fire Survey.

6. School properties will be inspected to ascertain that apparatus accessibility to the exterior portions of the school is insured. Where appropriate entrance to school yards is required for rescue and fire fighting purposes and is not provided, via a sufficiently wide entrance gate, a removable section of fencing must be maintained. This removable section will be provided with a center post painted red 'for easy identification. Inspecting members will ascertain that the required entrance to the school yards is not blocked and that appropriate "No Parking" signs are posted. In those cases where access to the school yard is not available, a duplicate memorandum will be forwarded to the Deputy Chief, Fire Prevention Division.

E.  RADIO COMMUNICATIONS

1. When leaving the station, the FCC will be notified via fire radio, that the company is going on block inspection. When inspections are performed by individuals and the apparatus is not t&en, the FCC will be notified, via fire phone, when leaving and returning to station.

# Appendix F (continued)

OPERATIONAL
PROCEDURE 4
AUGUST, 1990

2.    Upon arriving on location, where block inspections will be performed, the FCC will be notified. Whenever the apparatus is moved to a new location, the FCC will again be notified. This will serve to locate "dead" spots with regard to the fire radio and, additionally, inform other companies of your location.

3.    While a company is inspecting, the apparatus driver will remain with the apparatus to receive fire radio messages. If the company receives a run or is instructed to return to their station for any reason, the member at the apparatus will sound the siren to recall the other members. Members performing inspections will ensure that they are not too distant from the apparatus to hear the siren.

4.    On those apparatus where it is possible, the fire radio will be turned to the standby position and the engine shut off. This will conserve fuel and still enable the member remaining at the apparatus to monitor radio messages. If a radio transmission is required, the motor will be started before transmissions are made.

5.    Companies responding to an alarm while on block inspection will, upon completion of their assignment:

   a.    If a fire report is required, return 'to station, and complete same, and then resume block inspections. If Only an Analysis of Fire Alarm Resort is required, company will prepare the form after block inspections are completed for the day.

   b.    If no fire report is required, notify Fire Dispatcher that the company is resuming block inspection (give location).

6.    Companies dispatched to alarms during the hours of block inspections will use extra caution when responding, as companies on inspections may not be using their normal response routes.

# Appendix F (continued)

F.   REPORTS AND RECORDS

1.   Policies and Procedures

A folder will be maintained containing statements of Department, battalion or company policies and procedures concerning block inspections.

2.   Block Check Folders

A folder will be prepared for each block shown on the company local district map.  A drawing will be made on the front of each folder showing all streets contained in that block.  Folders will be maintained in numerical order.  Fire Prevention Check Forms (#76-24) will be retained, for each property, in the appropriate folder.

3.   Referral Reports

An "Action Pending" folder will be established for retention of Fire Inspection Referral Reports (#76-40) which have been forwarded by the company and on which no reply has been received from the Department of Licenses and Inspections via the Fire Marshal's Office.

All fire inspection referral forms returned to the company, from the Department of Licenses and Inspections, will be permanently retained in the appropriate block check folders.

4.   Board of Safety and Fire Prevention Reports

After a request for variance from the Fire Code is ruled on, a copy of the ruling will be sent through channels to the first-in engine company.  This copy of the variance ruling will be attached to the inside cover of the block folder and retained permanently and referred to when inspecting the property each year.  The date of the variance will be noted in the left hand margin.

5.   Progress Reports

Companies will record the number of regular and referral inspections performed each day on the office desk calendar.

# Appendix F (continued)

OPERATIONAL
PROCEDURE 4
AUGUST, 1990

If no inspections are performed that day, the reason
will be noted. .This information will be transferred
to the Company Block Check Progress Report and forwarded to the Battalion Chief on Saturday morning.
The Battalion Chief will consolidate the figures on
the Battalion Block Check Progress Report. Battalion
Chiefs will also call these figures into the Deputy
Chief of their division on Sunday night. When a
company/platoon has completed their block inspections, memorandum stating this will be
forwarded through' channels, to the Deputy Chief,
Fire Prevention Division. The Fire Prevention Division will maintain a record of completions.

V.  FORMS  REQUIRED

A.    FORMS

Consult the Forms Directive for preparation of all forms.

1.  Battalion Block Check Progress Report, #76-105.

2.  Company Block Check progress Report, #76-104.

3.  Complaint Report by City Employee, #70-35.

4.  Emergency Guide to Hazardous Materials Storage,
    #76-112.

5.  Fire Inspection Referral, S76-40.

6.  Fire Prevention Check, #76-24.

7.  Vital Building Information #76-80.

BY ORDER OF THE FIRE COMMISSIONER

# APPENDIX G

# Philadelphia Fire Code Amendments Adopted after the One Meridian Plaza Fire

(Bill No. 1466)

AN ORDINANCE

Explanation: [Brackets] indicate matter deleted.
Italics indicate new matter added.

Amending Chapter 5-1100 of the Fire Code, entitled "Fire Alarm Systems," by adding new sections regulating the maintenance of fire alarm systems, requiring notification of alarms and providing penalties; amending Chapter 5-1400 entitled "Fire Extinguishing Equipment by adding a new section relating to fire department connections; and amending Chapter 5-3 400, entitled 'High-Rise Buildings,* by adding new provisions to require all high-rise buildings, including existing buildings, to be equipped with automatic sprinkler systems and to meet other fire prevention and safety requirements set forth in Title 4, entitled "Building Code."

*The Council of the City of Philadelphia hereby ordains:*
Section 1. Title 5 of the Philadelphia Code, entitled "fire code," is hereby amended to read as follows

TITLE 5. FIRE CODE

# Appendix G (continued)

## CHAPTER 5-1100. FIRE ALARM SYSTEMS

* * *

§5-1104 Existing Systems

**§5-1105. Equipment to be Operative.**

*(1) Fire alarm systems and any other protective signaling systems which have been installedin compliance with any permit or order, or because of any law or ordinance, shall be maintained in operative condition at all times, and no owner or occupant shall reduce the effectiveness of the protection furnished.*

*(2) The provisions of §5-1105(1) shall not prohibit the owner or occupant from temporarily reducing or discontinuing the protection where necessary to make tests, repairs, alterations or additions. The Department shall be notified immediately by the owner, tenant, occupant, and/or operator of a supervisory service of the reduction or discontinuance of protection before such tests, repairs, alterations or additions are started. This notification to the Department shall set forth:*

*(a) the nature of the reduction or discontinuance of protection and the reason for it;*

*(b) the action being taken to make the tests, repairs, alterations or additions; and*

APP NO 59-3

*(c) the estimated date for restoration of protection. The Department shall be notified immediately upon the restoration of protection.*

*(3) No person shall install, repair or service any fire alarm or protective signaling system unless that person has adequate knowledge of the operation and handling ( such equipment. The Department shall by regulation prescribe minimum experience and other requirements for such persons.*

*(4) Any person servicing any fire alarm or protection signaling system shall attach thereon a card showing the date when such work is done, and the name, address, an telephone number of the person or firm performing such work.*

*(5) Fire alarm and other protective signaling system shall be inspected at least annually by qualified person, and inspection records maintained by the owner, tenant or operator of the building.*

**§5-1106. Notification to Fire Department.**

*(1) The Department shall be notified immediately the owner, tenant, occupant, operator of a supervision service, and/or a central station service of the activation of any fire alarm signal.*

*(a) Alarm signals initiated by manual fire ala boxes, automatic fire detectors, waterflow alarms for,*

# Appendix G (continued)

APP. NO. 594-4

automatic sprinkler systems or activation of other fire suppression systems or equipment shall be treated as fire alarms.

(b) Central station services shall immediately retransmit alarm signals received from protected properties to the Department.

[§5-1105]§5-1107. Penalties.

• • •

(2) In addition to any other penalties provided in this Title any person who violates §5-1106(1) or the regulations issued thereunder, shall be subject to a fine of three hundred ($300.00) dollars for each violation or to imprisonment not exceeding ninety (90) days, or both.

(3) In addition to any other procedures, remedies or penalties provided by this Title, the provisions of §5-1106(1) may be enforced in the manner set forth in this section.

(a) Notice of violation.

(1) Whenever an investigation by the Department of licenses and inspections or the Department discloses any violation of the provisions of §5-1106(1), the violator shall be issued a printed notice of violation by the person conducting the inspection. Such notice shall bear the date, time and nature of the violation, identity and address of the violator, the amount to be remitted in response to the

APP. NO. 594-5

notice of violation, and the penalty which can be imposed by the court for the violation. The notice shall be signed by the person issuing the notice and shall bear the official identification number of the person issuing the violation notice. For each violation, a separate notice of violation may be issued under this section.

(2) Any person who receives a notice of violation, may within ten (10) days, pay the amount of three hundred ($300.00) dollars, admit the violation and waive appearance before a Municipal Court Judge. The notice of violation shall contain an appropriate statement for signature by the violator for the purpose of admitting the violation when he or she remits the stipulated payment. Any payment made under this subsection shall not relieve the violator of the responsibility for correcting the violation set forth in the notice of violation.

(3) If a person who receives a notice of violation fails to pay the prescribed payment within ten (10) days of the issuance of the notice of violation, a code enforcement complaint shall be issued for the violation in such a manner provided by law. If the person named in the code enforcement complaint is found to have violated §5-1106(1), he or she shall be subject to the penalties set forth in §5-1107(2).

(4) Any fine or costs imposed by the courts shall be entered as a judgment against the violator. Any fine and

# Appendix G (continued)

APP NO. 594-6

costs imposed by the court shall be paid within ten (10) days of its imposition. If the fine together with any court cost is not paid within such period, the violator shall be subject to proceedings for contempt of court and/or collection of the fine as provided by law.

(4) In addition to any other sanction or remedial procedure provided in this Title, any person who shall violate the provisions of §5-110611) or the regulations pursuant thereto, shall be subject to referral to the Office of the District Attorney.

    •
        •
    •
        •

CHAPTER 5-1400. FIRE EXTINGUISHING EQUIPMENT

§5-1403.  Equipment to be Operative.

(1) Sprinkler systems, standpipe systems, [alarm systems,] and any -other [protective or] extinguishing systems which have been installed in compliance with any permit or order, or because of any law or ordinance, shall be maintained in operative conditions at all times, and no owner or occupant shall reduce the effectiveness of the protection furnished.

    •
        •

§5-1405. Fire Department Connections on Sprinkler and Standpipe Systems.

APP NO. 594-7

(1) All fire department intake connections to sprinkler and standpipe systems shall have two (2) or more two an one-half inch female National Hose (NH) standard three intake fittings. If an intake connection feeds one (1) zon rather than the entire building, the floors encompassed . the zone shall be indicated 'at the intake which feeds 11 zone.

Exception: Dry standpipe systems with six inch riser piping shall have connections with three two and one-half inch female NH standard thread intake fittings.

(2) Standpipe outlets on Class I and Class III system that is, those designed for fire department use, shall have floor outlets with two and one-half inch male NH standard thread fittings.

    •
        •
    •
        •

CHAPTER 5-3400. HIGH-RISE BUILDINGS.

§5-3402.  Aplication.

(1) [New high-rise buildings shall conform to Sect. 629-0 "High-Rise Buildings" of the Basic Building Code/1981 published by the Building Officials and Code Administrators (BOCA) International, Inc. and to applicable provisions of this Chapter.]

# Appendix G (continued)

NO. 594-8

Existing -high-rise buildings shall conform to 602.3 'Sprinkler system' of the Building Officials Code Administrators (BOCA) National Building Code Eleventh Edition, 1990, as incorporated by Title 4 of Philadelphia Code, entitled 'Building Code,' and to provisions of this Chapter.

Exception: In R-2 occupancies, as defined in the Building Code sprinklers are only required in areas as specified by Section 5-3405 of this Code provided that:

(1) Smoke detectors connected to the building's electric system are provided in each dwelling unit and installed in accordance with Section 1018 of the Building Code within two (2) years of the date of this chapter.

Exception: Electrical interconnection of smoke detectors within a dwelling unit is not required, provided that the alarm level of each detector is 15 decibels above the ambient noise level in every occupied space within the unit.

(2) Doors from dwelling units opening into a common corridor are self-closing within one (1) year of the effective date of this chapter.

within one (1) year of the effective date of this Chapter the of every existing high-rise building shall submit to

APP. NO. 594-9

the Department of Licenses and Inspections for approval a detailed written description of the methods and schedule to be used for compliance with this Chapter. In no case shall the plan provide for the completion of all work required by this section later than eight (8) years from the effective date of this Chapter. Applications for permits necessary for compliance with the provisions of this Chapter shall be submitted to the Department of Licenses and Inspections upon approval of the written description in accordance with the following schedule:

(1) Within eighteen (18) months, water supplies to all floors of the building, and compliance with the sprinkler and standpipe connection requirements in §5-1405.

(2) Within thirty (30) months, fire suppression systems in twenty percent (20%) of the floors of the building.

(3) Within thirty-nine (39) months, fire suppression systems in forty percent (40%) of the floors of the building;

(4) Within forty-eight (48) months, fire suppression systems in sixty percent (60%) of the floors of the building;

(5) Within fifty-seven (57) months, fire suppression systems in eighty percent (80%) of the floors of the building; and

# Appendix G (continued)

APP. NO. 594-11

Inspections. Tests shall include 11 transfer from normal to standby and/or emergency power under full load operating conditions. Tests shall be performed by a licensed electrician who shall certify to the Department of Licenses and Inspections that the system is in proper working order. Any failures shall also be reported and shall be repaired and the system retested within thirty (30) days.

§5-3410 Notification of Fire Department.

The Department shall be notified immediately of any changes to the fire suppression, fire alarm, emergency electrical and any other building systems which are necessary during fire department operations in a building. Detailed information and instructions on all equipment associated with these systems shall be provided to the Department.

SECTION 2. This Ordinance shall take effect immediately

Explanation
[Bracket] indicate matter deleted
Italics indicate new matter added

NO. 59.1.10

(b) Within sixty-six (66) months, fire suppression as in one hundred percent (100%) of the floors of the building

to the expiration of one (1) year after the effective of this Chapter.] Upon notification by the department of Licenses and Inspections that each permit fee is complete, the applicant shall obtain permits thirty (30) days [of notification]

•

•

•

3105 Fire Suppression Systems

All existing high-rise buildings where standpipe 1.5 exceed two hundred seventy-five (275) feet from the at fire department connection to the roof or highest action shall be equipped with wet standpipe systems called in accordance with NFPA 14 and the Building Applications for permits necessary for compliance the provisions of this section shall be shall be submitted to the department of Licenses and Inspections within three (3) of the effective date of this chapter

•

•

•

199 Testing of Electrical Systems

A standby and/or emergency electrical systems shall be tested at least annually in accordance with the action of the Department of Licenses and

# Appendix G (continued)

APP. NO. 594-12

CERTIFICATION: This is a true and correct copy of the original Ordinance approved by the Mayor on

**DECEMBER 18,1991**

Deputy Chief Clerk of the Council

# APPENDIX H

# Photographs

Numerous slides and photographs are included with the master report at the United States Fire Administration. The photographs presented on the following pages were taken by Charles Jennings after the fire, except where otherwise noted.

## Appendix H (continued)

*Philadelphia Inquirer photo by Michael S. Wirtz*

Exterior view of One Meridian Plaza and fireground operations in the early morning hours of February 24, 1991. Fire involves the 22nd, 23rd, 24th, and part of the 25th floors. Note the heavy stream played on the exterior from an adjacent building.

# Appendix H (continued)

*Philadelphia Inquirer photo by Michael Mally*

Aerial view of exterior firefighting operations after dawn on February 24, 1991.

# Appendix H (continued)

*Philadelphia Inquirer photo by Michael Mally*

Exterior firefighting efforts from the Girard Building #1, east of One Meridian Plaza. The two buildings are connected on lower floors.

# Appendix H (continued)

*Philadelphia Inquirer photo by Rick Bowmer*

Smoke pours from One Meridian Plaza as William Penn looks on from atop Philadelphia City Hall the morning of February 24, 1991, after interior firefighting efforts have been suspended.

# Appendix H (continued)

*Philadelphia Inquirer photo by Michael Mally*

**View of deluge set operating from One Centre Square.**

# Appendix H (continued)

**Exterior view of building looking south from City Hall Plaza.**

# Appendix H (continued)

Exterior view of south side of building.

## Appendix H (continued)

This stairway connected the One Meridian Plaza Building with the adjacent office building.

# Appendix H (continued)

Here and in the next two photos exterior granite panels from the east stair tower were dislodged due to the thermal expansion of the steel frame of the building.

# Appendix H (continued)

# Appendix H (continued)

# Appendix H (continued)

Photo by James David

One of the areas of fire penetration on the 30th floor where a single sprinkler head activated to stop the upward extension of the fire.

# Appendix H (continued)

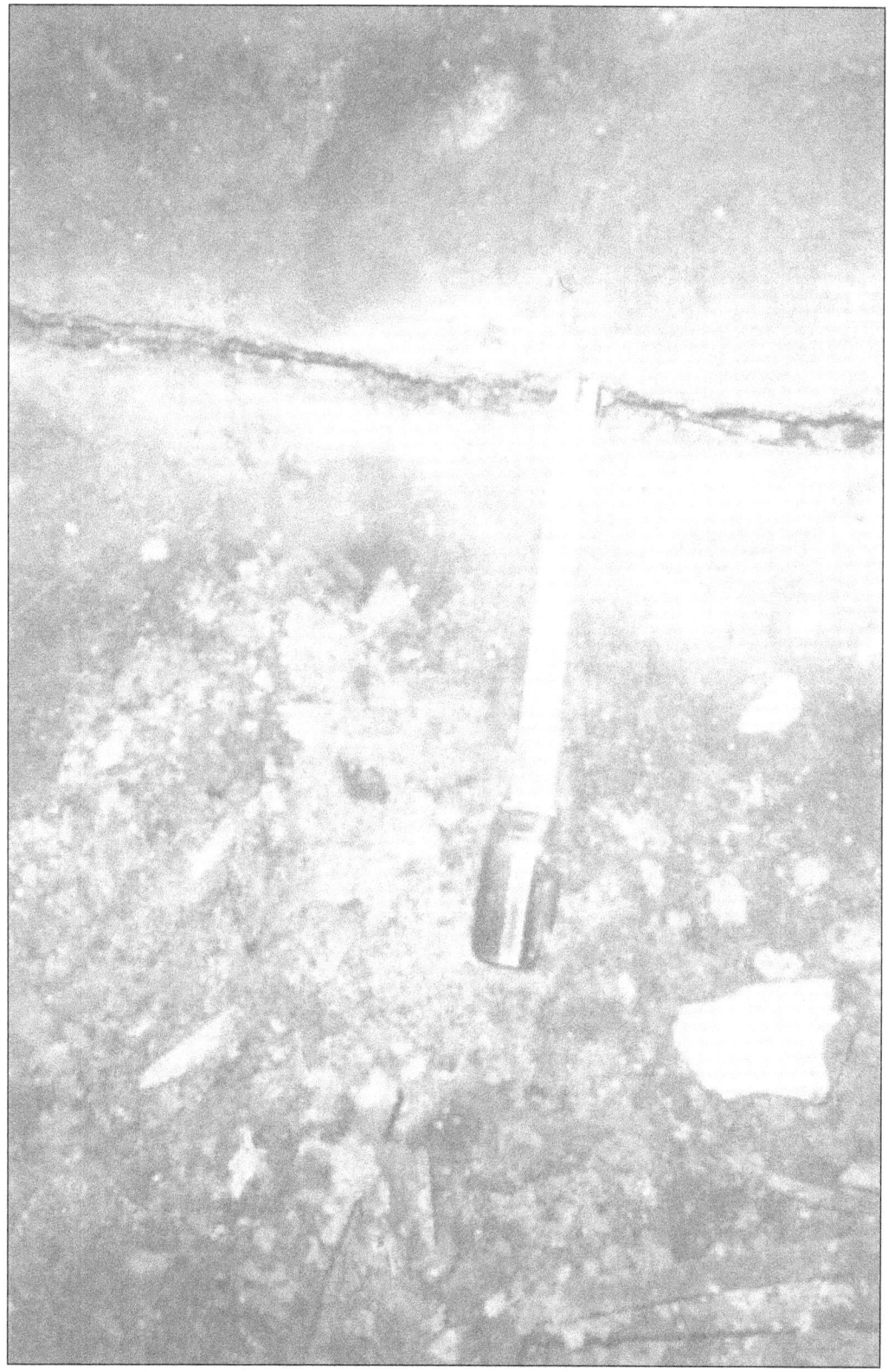

Close-up of crack in concrete floor, 28th floor.

## Appendix H (continued)

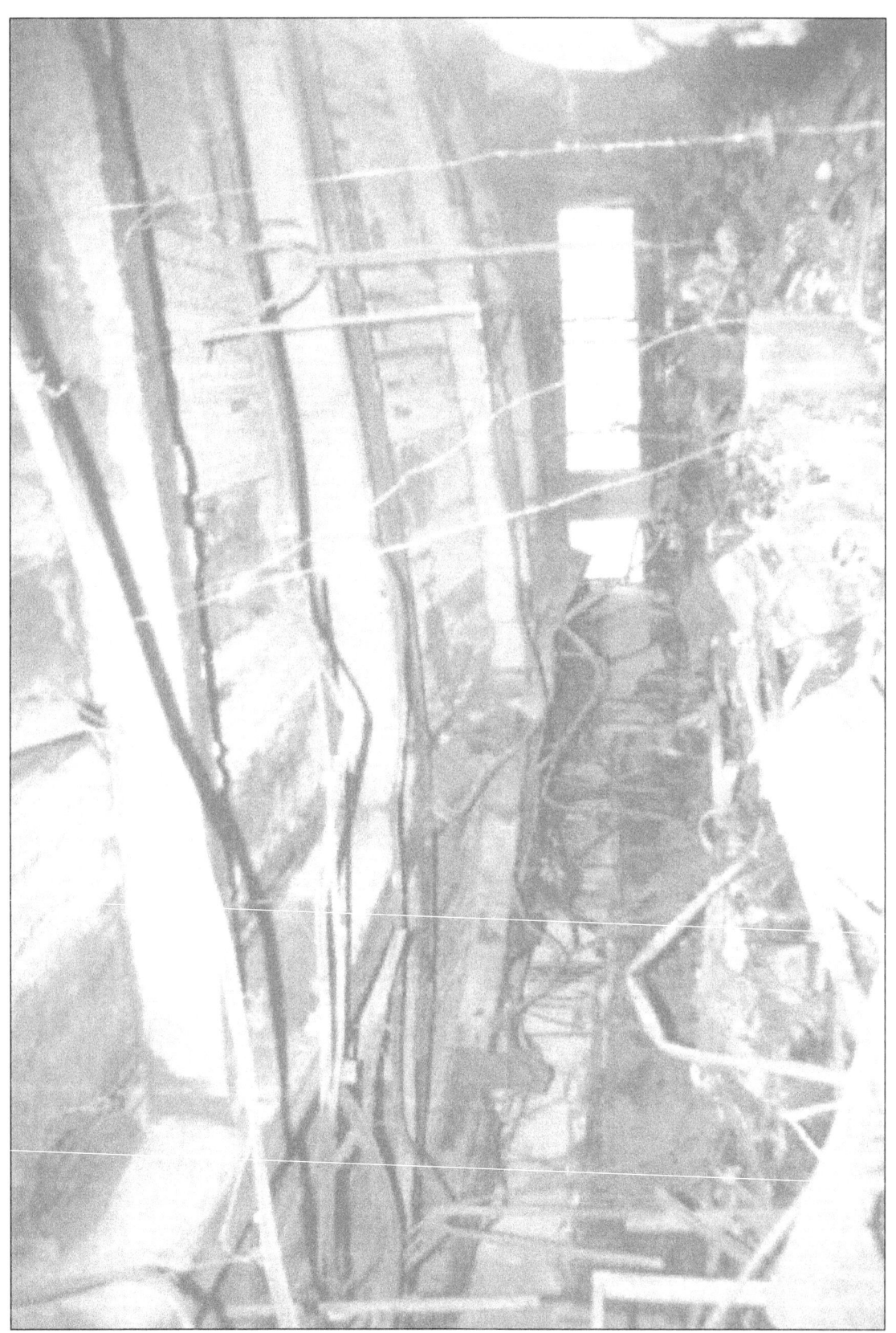

Here and in the next three photos are interior views of floor areas after the fire. Note the total consumption of the available fuel and sagging of the floor deck of up to three feet between columns.

# Appendix H (continued)

# Appendix H (continued)

# Appendix H (continued)

# Appendix H (continued)

Detail of typical Type 1B construction, 6th floor.  Note the spray-applied fireproofing and framing for gypsum wallboard.

# Appendix H (continued)

Fire tower between 25th and 26th floors.  Note heavy fire and smoke damage.

# Appendix H (continued)

Occupant use standpipe hose cabinet on 26th floor.

# Appendix H (continued)

Standpipe hose outlet with pressure reducing valve (PRV), 26th floor.

# Appendix H (continued)

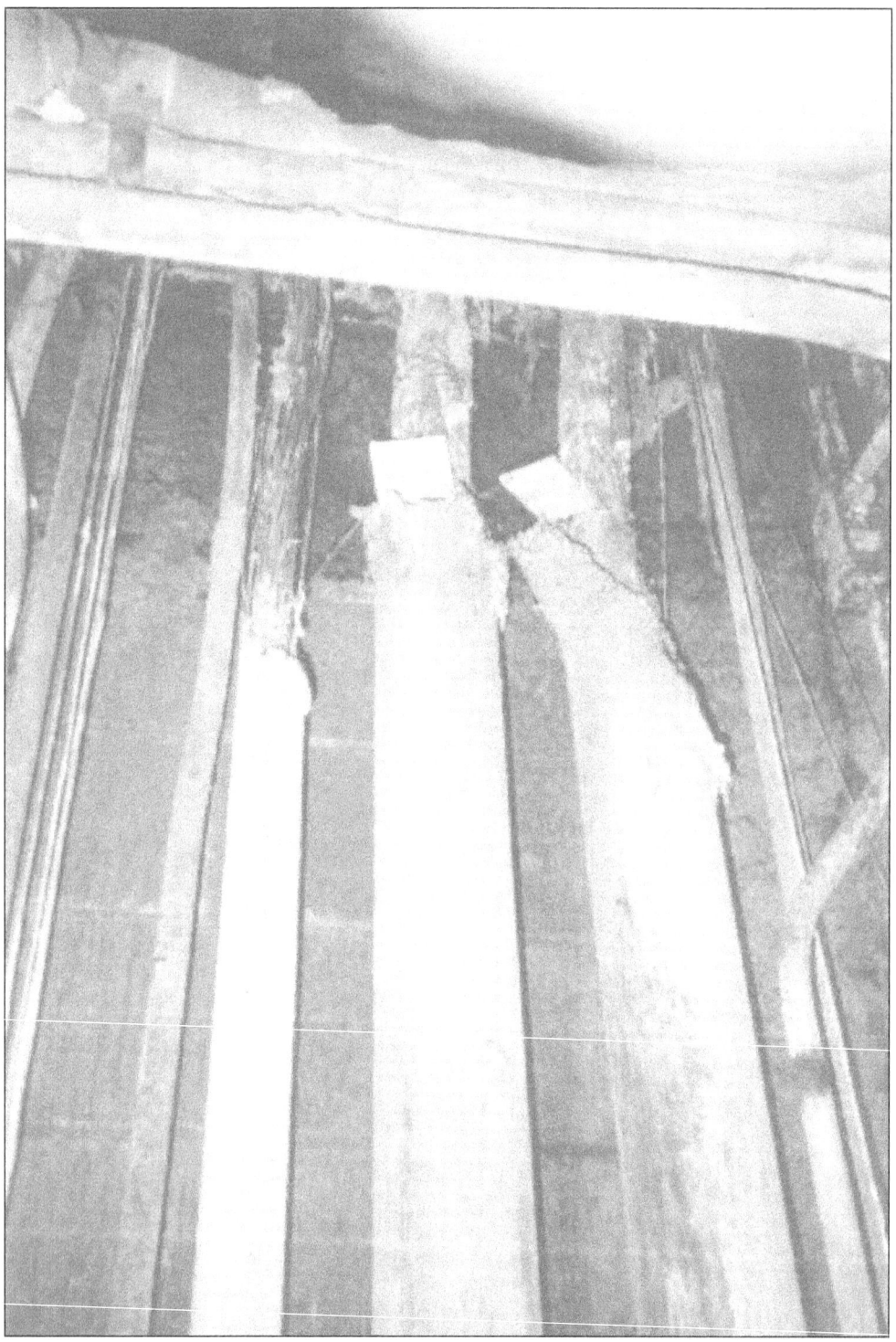

Electrical shaft enclosure on the 30th floor, showing side-by-side risers for the two power supplies, both damaged by fire penetration at the plenum level of the adjoining office space.

## Appendix H (continued)

Section of 5-inch hose that was ruptured by falling debris outside the building. Hose lines feeding the standpipes and entering the stairways were damaged several times and had to be replaced at great risk to firefighters. (Shoring boards were later used to protect the hose lines.)

# Appendix H (continued)

Street in front of the building from front steps showing stranded autos and debris in street.

# Appendix H (continued)

Rooftop heliport.

www.ingramcontent.com/pod-product-compliance
Lightning Source LLC
Chambersburg PA
CBHW081142170526
45165CB00008B/2764